Samsara

An Exploration of the Hidden Forces that Shape and Bind Us

Samsara

An Exploration of the Hidden Forces
that Shape and Bind Us

Daniel McKenzie

MANTRA
BOOKS

Winchester, UK
Washington, USA

JOHN HUNT PUBLISHING

First published by Mantra Books, 2021
Mantra Books is an imprint of John Hunt Publishing Ltd., No. 3 East Street, Alresford
Hampshire SO24 9EE, UK
office@jhpbooks.com
www.johnhuntpublishing.com
www.mantra-books.net

For distributor details and how to order please visit the 'Ordering' section on our website.

Text copyright: Daniel McKenzie 2020

ISBN: 978 1 78904 894 0
978 1 78904 895 7 (ebook)
Library of Congress Control Number: 2021933144

A CIP catalogue record for this book is available from the British Library.

Design: Stuart Davies

UK: Printed and bound by CPI Group (UK) Ltd, Croydon, CR0 4YY
Printed in North America by CPI GPS partners

We operate a distinctive and ethical publishing philosophy in
all areas of our business, from our global network of authors to
production and worldwide distribution.

Contents

Also by Daniel McKenzie

The Wisdom Teachings of the Bhagavad Gita: The Secret to a Life Free of Suffering
ASIN: B087MXW7L8

The Broken Tusk: Seeing Through the Lens of Vedanta
ASIN: B07KS2WVRN

Essays by Daniel McKenzie: www.TheBrokenTusk.com

Foreword

Life is pleasure and life is pain. It is joy and sorrow woven fine. For most of us, how it works and how to live successfully is a great mystery. Ever since humans started walking upright, the knowledge of their environment, who they are and the mysterious factor that created life began to accumulate and organize itself into an elegant, logical, useful science—Vedanta—that solves the mystery without removing the wonder. The difficulty involved in successfully negotiating life's swirling, unpredictable, sometimes dangerous currents is due to ignorance of the nature of the world and the nature of our own hearts and minds. The ancients called this difficulty *samsara*. Samsara is a frustrating zero-sum state of mind that cries loudly for a solution, particularly so in these trying times. As our environment degrades, our minds fray. We question life and we question ourselves all the while generating desperate and/or comforting counterproductive stories rooted in lack as we struggle to understand.

Yet we needn't worry. Daniel McKenzie in *Samsara: An Exploration of the Hidden Forces that Shape and Bind Us*, informs us that the secret—a well-travelled path leading out of the mind's dark *samsaric* prison into the blissful light of freedom—is revealed in the essence of mankind's commonsense science of life, the Bhagavad Gita, which appeared on the scene several centuries before the Christian Era and is as applicable today as it has been over the centuries.

The remedy, relearning what it is to be a human being, which requires patience and honesty, is a three-step process based on knowledge generated by using our mind's inbuilt power of discrimination and the heart's determination to remove the beliefs and opinions, dreams and fantasies that keep us enslaved to fears and desires too numerous to mention. It is a journey of discovery well worth the effort.

—James Swartz, author of *The Essence of Enlightenment: Vedanta, the Science of Consciousness*

Preface

The interest in writing a book on *samsara* came after meditating for some time on the "field" in which we work, play and experience life. If you pay attention to the media, you may be persuaded to believe that life is mostly dominated by chance, disorder and chaos. But look closely and you will find a world operating in perfect order.

Ask any scientist and they will tell you our world is one full of observable patterns, connections and laws and it's only because of those reliable patterns, connections and laws that we are able to formulate certain outcomes and even define what order is. While things may sometimes appear messy and untidy, it's the self-regulating nature of the world to eventually plane and make flat that which is uneven.

Like everything else, human beings are a product of the world and as a result, are formed and influenced by the same patterns, connections and laws as the rest of nature. However, due to our intellects there are additional psychological and moral laws that apply only to us. If we contradict those laws, we pay the price in unpleasant ways; for example, with guilt, addiction, anxiety and a whole litany of maladies all too common in the twenty-first century.

Samsara is often associated with worldliness, or the continuous cycle of birth and death where souls travel from one incarnation to the next in pursuit of objects and experiences that promise to fulfill them. But perhaps samsara is better understood to be the psychological realm within which human beings operate. This realm or system is built on the principle of cause and effect and delivering the results of one's actions in a way that either rewards or discourages them. This unique form of governing protects and sustains the total and helps ensure everything doesn't suddenly come to an abrupt end.

The more I thought about this intelligent, self-regulating system, the more I realized that life is a setup—a kind of elaborate amusement park perfectly built to frustrate and bring us closer to the truth. The system not only teaches us in direct and indirect ways to, for example, not play with fire or steal from our neighbor, but via our mistakes, misapprehensions, and suffering, it also makes us inquisitive. When we experience some pain it's normal to stop and ask ourselves, *What just happened, why am I so unhappy?*

Needless to say, the system comes with a steep learning curve. Most people learn how to navigate life via "The School of Hard Knocks"—that is, by whatever experience dictates in negative ways. Our education is not so much derived from direct knowledge of universal rules, but from the gradual, albeit painful education that comes from falling down over and over again. With such a blunt approach to learning, the best we can hope for is that by repeatedly scraping our knees, our behavior will eventually change, and we will grow wiser.

The system can also feel like it's constantly undermining our freedom. For every up, there's a down; for every down, an up. We can't win! Add to that the fact that we find only temporary satisfaction in the objects, relationships and experiences we pursue, and one can begin to see the system is purposely built to exasperate.

We travel through life (perhaps, lives), crossing the desert of the world chasing one mirage after another in hopes that it will finally fulfill us, until one day we stop. And instead of looking out toward the horizon at the next shiny object, we begin to look within. Only then, do we begin to awaken from the dream that is samsara.

In writing this book, it is my attempt to show the reader that the concept of samsara is much more than just a passing metaphor for the soul's endless transmigration and fruitless pursuit of happiness. Samsara is a condition we're all unconsciously trying to escape from. And yet, like Arjuna, the hero in the Bhagavad Gita, before engaging in battle we must first understand what we're up against. Only then, do we have a fighting chance of obtaining the knowledge necessary to set us yonder.

Daniel McKenzie

February, 2021

Chapter 1

What Is Samsara?

In popular culture, samsara is often portrayed as something exotic, sensual or pleasure-inducing. It's no wonder its name has been used to sell everything from lady's perfume to herbal supplements. These days, we even find it used to promote technology. The website, Samsara.com sells internet-of-things sensor models, describing their product as being able to "securely connect sensor data to the Samsara cloud."

We can assume samsara was also the inspiration behind the 1999 blockbuster hit, *The Matrix* which converted the concept into a simulated reality where intelligent machines have taken control over the minds of humans, distracting them in order to use their bodies as a source of energy. The problem with *The Matrix* story is that although the leading character escapes the samsara created by the machines, he never questions the samsara in where he now plays the role of "the chosen one" who saves all of humanity. In other words, from the Vedic perspective, our hero is still asleep.

There is also the moving visual feast by the same name *Samsara* that came out in 2011. Described as a silent documentary, *Samsara* "transports us to the varied worlds of sacred grounds, disaster zones, industrial complexes, and natural wonders." Actually, *Samsara* (the movie) does a pretty good job of showing the dual nature of samsara (the concept). A prominent feature of samsara (the concept) is its pairs of opposites, including both its splendid beauty and awesome destruction. As it turns out, samsara is indeed, a very strange place.

The word samsara comes from Sanskrit, meaning to "flow together," which alludes to the flux and flow of the universe and empirical existence. In spite of the many sensual delights, samsara has to offer, within eastern spiritual traditions such as Buddhism and Hinduism the term is almost exclusively associated with being bound or limited. In these traditions, life is not portrayed as an endless playground for reaping pleasure, but instead, something more akin to imprisonment where individuals work out the effects of their past deeds in order to eventually obtain liberation.

In the Vedic tradition from which Hinduism is based, samsara is typically taught using the example of the warrior-prince, Arjuna, from the Bhagavad Gita. In the Gita (a subset of the Indian epic, *Mahabharata)* there is a dialog between Arjuna and his charioteer Krishna before a great righteous war. Taking inventory of the battlefield, Arjuna comes to the unsettling realization that in order to win the battle, he must set out to destroy his own kin and beloved teachers who have taken sides with a ruthless demagogue. Due to this predicament, Arjuna is overtaken by the classic signs of samsara, which include *attachment, despair* and *delusion.*

The entire first chapter of the Gita is an exposition of Arjuna's attachment and grief culminating in him hopelessly throwing down his weapon. It's only after Arjuna's visible breakdown that Krishna accepts the role as *guru* and begins to methodically show Arjuna the way out of his confusion and back to defending the social order. Thus, the Gita begins with dramatic force, showing the reader how attachment and our inability to see reality for what it is, binds and keeps us suffering in samsara. Through Arjuna's dilemma and tribulation we learn that samsara is not something "out there" but instead, a condition within the mind rooted in ignorance.

Ignorance, in this case, doesn't mean "stupid" but instead suggests a sort of blind spot where one is unable to see the truth. The noble prince is confused about his moral responsibility to defend *dharma*—the universal laws that keep society together. He is unable to look past his fondness for his family, friends and teachers in order to do what's in the best interests of the total.

Krishna reminds Arjuna that he isn't seeing reality for what it is and that he needs to step back for a moment and consider that "the wise grieve neither for the living or for the dead." Krishna then uses the remaining sixteen chapters of the Gita to unpack this enigmatic assertion and in the process, show Arjuna he has nothing to fear. But even without yet understanding this opening verse of the Gita, we're able to define samsara as:

A negative psychological condition brought on by the misinterpretation of reality.

As we will soon discover, samsara has many related meanings, but they are all derived from the idea that we suffer because we're unable to see the truth about reality. In short, to paraphrase Krishna, we grieve for that which needn't be grieved for.

But you don't need to read the entire *Mahabharata* in order to get a handle on samsara. We can already witness its negative effect in our own life and in those of others. The ancient texts tell us we go through life constantly chasing objects, relationships and experiences due to our own ignorance. So already, one might get a sense that samsara isn't something you want, it's something you want to get out of!

One of the most common ways we are led into the trap of samsara is via the belief that joy is in the object. The book and documentary, *Generation Wealth* by social anthropologist, photographer and director Lauren Greenfield makes the point by showing extreme cases of both poor and rich stuck in the mud of samsara. In her 2018 documentary, Greenfield has checked in on several people over a multi-year span who find themselves waist-deep in it. She interviews a former hedge-fund manager living in exile who is on the "FBI's Most Wanted" for financial fraud. At one time worth approximately $800 million and working 100 hours per week, he believed happiness was money. Greenfield interviews a young porn star who took her profession to the extreme before waking up to the pedophile fantasy she was perpetuating and the abuse she was causing to her body. She believed happiness was easy money and being a celebrity. Greenfield also talks to a woman who on bus driver wages and a loan from her mother, travels to Brazil to have extensive cosmetic surgery. In spite of a family tragedy related to her obsession with her physical appearance, she continues to still believe that happiness is the body.

Generation Wealth paints such an accurate picture of samsara that after watching the documentary, you can't help but be convinced the world is nothing but an elaborate setup to thwart us from achieving our goals of gaining total satisfaction. One thing those interviewed all have in common is that each had to hit rock bottom before trying to get out; each one had to suffer immense personal losses from their delusion before finally waking up and looking for an exit.

In case there was any doubt, delusion is not a happy state. When we are deluded, we are under the spell of *maya*—samsara's means of projecting the false and concealing the truth. Nobody is immune to maya's powers, even the most enlightened can still fall victim from time to time. Samsara's grasp is so strong that it is often likened to being caught in the jaws of a crocodile. Here, delusion is the belief that by doing more of the same, I will eventually be satisfied. It's like the alcoholic promising to quit after one more drink. The problem is samsara has no end, it's a bottomless pit that goes on and on, slowly enveloping us and leading to self-entrapment.

Upon analysis, we can identify three outcomes of pursuing samsara and the related belief that joy lies in objects. The first one is obvious: suffering. The wise say that in life pain is inevitable, but suffering is optional. Mostly, we suffer because we can't get what we want. This theme is repeated over and over again ad nauseam in popular culture, especially in popular music whose lyrics typically include some variation of "guy gets girl, guy loses girl."

The next outcome is dissatisfaction. What we desire and are finally able to obtain cannot provide us with the permanent joy we seek. Due to the changing nature of the field of experience, all objects have an expiration date where the initial thrill dissipates and eventually disappears. In samsara, the first time really is the best! Thereafter, an object's joy is like a depleting battery. This inevitable loss quickly leads to a feeling of poverty and even anger as we try to repeat the same pleasure. Thus, once the spell is broken and the joy gone, we must seek another object to pursue. The samsari's mantra is always "more, better, different" because that's what's needed to externalize happiness and sustain the belief that joy is in objects.

All objects are ultimately disappointing because all objects are constantly changing and nothing in samsara is ever what it seems. Examined closely, we find that objects are not only constantly in the process of becoming something else but are made up of other smaller parts or aggregates that are divisible. That objects appear substantial is only an illusion. Using the concept of a shirt, if we take away the form "shirt," what we are left with is fabric; take away the fabric and we have cotton; take away the cotton and we have fiber; take away the fiber... and so it goes all the way down to the molecular level. What we believe to be concrete and permanent is really just a name and form we apply to a heap of other stuff. In fact, upon close inspection all objects dissolve into seemingly, nothing. This is why the idea that by clinging to a certain object (person, place or thing) we will gain ever-lasting joy, is such a losing proposition. How do you cling to something that lacks substance and is forever turning into something else? It's no wonder we suffer.

Lastly, is dependence. In order to feel the initial thrill an object gave us the first time, we must either increase the amount we're exposed to or increase the frequency at which we experience it. Unfortunately, this is a recipe for addiction, which obviously leads to a myriad of health issues. And as we all know, addiction is not happiness; addiction is being in the jaws of the crocodile.

All of this isn't to say we should never enjoy objects, relationships and experiences. Like a magnificent museum that delights and enchants, we are here to enjoy life with all its variety and wonders. Just don't think you can walk away with any of the exhibits. Experience can only ever be enjoyed as a fleeting moment.

But to cut to the chase, joy isn't in objects, *it's in you!*

If joy were in objects, the same object that gives you happiness would give me happiness too. The fact that we continue to believe joy exists in inert objects even after learning that it doesn't is one of samsara's tricks of reversal. As we'll learn later, the joy we experience from obtaining an object of our desire isn't the result of obtaining it, but rather from not desiring it anymore. Desire is like an itch that we create with our thoughts. Of course, the only way to satisfy an itch is by scratching it, which only becomes a problem when, like a dog taken over by fleas, we find ourselves scratching all the time.

We've all tasted samsara and have unknowingly trapped ourselves in it in little and big ways. We all experiment with rolling around in the mud for a while especially when we're young and lack the wisdom experience provides. Society is constantly promoting a samsaric lifestyle with its relentless advertisements, pervasive media, and consumer-focused holidays. Advertising, in particular, is such a powerful force because it promotes a feeling of poverty. Through repetition we are brainwashed to want things we didn't even know we needed! The result is a world obsessed with constantly trying to improve itself via gadgets, diet fads, personal trainers, self-help gurus, cosmetic surgeons, and more.

For some, it's because of this feeling of lack that they begin to inquire. They realize that no amount of pleasure, wealth, or recognition can ever make them satisfied. This feeling of limitation is rooted in a simple misunderstanding about who/what we are. If we believe we are the body and are in constant opposition to it getting old, weak and sick, we're always going to be disappointed because nobody can fend off time's inevitable course of destruction. And if we believe we are the ego and must consume experiences in order to make ourselves feel full and satisfied, we're always going to be hungry because all experience is only temporary. But if we understand that the essence of who we are is actually whole, limitless and changeless—in other words, if we understand that we're already full—with some work we will eventually lose the insecurities we have about plenty never being enough.

But if we aren't the body or the ego, then who or what are we?

We are that which is witness to all experience and objects. We are that which knew the baby it once was become an infant, turn into a teenager, and then an adult. We are that which has witnessed the person go to college, get a job, find a partner, get married, have children and perhaps divorced. This witness is that which never changes because if it did, we would never know that we were once a child.

You can't observe change if you are the change, just like you can't tell you're on a moving train if the train next to you is moving at the same speed. It's only by being still against a moving foreground that one can recognize change. And that's what we are—the stillness, awareness, a.k.a. the "Self."

In contrast to dream-like samsara—which includes all sentient beings and inert objects whose nature is to change—the Self is that which is real because the Self is that which is always present and immutable. Once we identify with the Self instead of the external changing objects we perceive, samsara begins to loosen its grip on us. Because if we are the changeless and limitless Self, then there is nothing more to add—we are already full! And while this truth may be difficult to grasp now, once realized it becomes our ticket to freedom.

So far, we can state the following about samsara:

- Samsara is a condition of the mind rooted in ignorance.
- Samsara is the belief that I am incomplete and that my happiness is dependent on objects.
- Samsara is entrapment through attachment that leads to sorrow and delusion.

And as an aside, as intriguing as it sounds, our present concern with samsara shouldn't be about past or future lives (at best, a belief) but how we suffer now, in this very life due to conditions and forces we barely understand, including the truth about who/what we are. Fortunately for us, there is a cure for samsara, there is a way out, and that way out is through knowledge. How else would we solve the problem of ignorance?

Chapter 2

Beautiful, Intelligent Ignorance

To better understand samsara we need to go directly to its source—ignorance. One can define ignorance as a lack of discriminative knowledge and yet, ignorance is more than just not knowing. Ignorance is very intelligent in the way it guards itself by projecting the world, filling the mind with desire and turning our gaze outward. Ignorance, for example, is what magicians rely on to trick their audience. Magicians look for blind spots, vulnerabilities and the limits of their audience's perception in order to influence them and create an illusion. Once the magician knows which buttons to push, they are able to play their audience like a piano. In the Bhagavad Gita, Krishna reveals to Arjuna that "The Lord lives in the heart of all beings. By its power to delude, it causes them to dance as if they were puppets on a string."

We are all born ignorant, it comes with the package of being human. As we mature, we begin to appreciate the value of knowledge and seek it more because being ignorant, for the most part, is painful. The old adage, "ignorance is bliss" might apply to children or those beings who operate based on their own innate programming, but it doesn't apply to anyone with a fully developed intellect. Nobody chooses to be ignorant. To be ignorant is equivalent to a life of confusion and error, forever repeating the same mistakes. After all, we all wish for more answers, not more questions.

To illustrate the idea, imagine you were born and raised in New York City and woke up one day to mysteriously find yourself in the jungle, alone. While navigating the streets of New York City might come easy to you, trying to navigate the jungle would most certainly not. And yet, to someone who was born and raised in the jungle, it would be no more threatening than a stroll down Fifth Avenue. The single most important factor needed to keep you alive in the jungle (or NYC, for that matter) is knowledge. No wishing or hoping is going to get you out of the jungle safely. Without knowledge, we are like a babe without its mother, left vulnerable to the winds of change.

As a society, we don't value knowledge enough, often preferring to side with our beliefs or opinions in a gamble that reality will somehow magically align with them. We choose to ignore scientists, journalists, accountants and other truth-finding experts in favor of a rosier outlook—one that doesn't crimp our style. History is replete with fools trying to subvert the facts in order to remodel the truth to their own liking.

Every day we are persuaded to believe in ideas that don't match with reality. Sometimes we think it through and decide what's being sold to us doesn't make sense and we put it aside. Other times, our wish that things be different from reality takes over and we willingly choose to look the other way. Lastly, sometimes we just don't know what we don't know and are led into a decision blindly.

Battling ignorance can be an uphill battle. The reason is because ignorance is hard-wired. One way it's hard-wired is through our already firmly established beliefs. Some of these beliefs were planted in us when we were very young, while others we have cultivated voluntarily in order to create a sense of hope or security. Ignorance is also hard-wired because we are unwilling to challenge the foundation for what has taken us a lifetime to build. Most of us will defend our beliefs tooth and nail before admitting they don't add up and that we are wrong.

There's also our conditioning. Due to habits, we've picked up over a lifetime, we have been programmed to behave in certain ways depending on the situation. Mostly, we fall for the same painful traps and delusion because that's the way we've always done it. Sadly, many of us constantly act out in ways that go against our own best interests. You know ignorance is deeply entrenched when it takes all your willpower to not succumb to a desire you know you'll later regret. "Just one more slice of pizza," we tell ourselves, or "This is my last cigarette," "drink," "sexual encounter," etc. It's at that point that you have to begin asking yourself, who is in charge—me or my desires? So, ignorance doesn't just work based on what we don't know, it can also work based on what we already are aware of, but choose to ignore—in other words, a willful ignorance of the facts.

How Ignorance Fools Us

Over and over again, due to its keen ability to project and conceal, we fall under the spell of ignorance. The spell isn't just that little voice that tells you, for example, it's okay to buy a new car you can't really afford. Ignorance permeates our entire experience and how we interpret the world, right down to the molecular level.

You can't talk about samsara without a mention of maya. Like samsara, maya is described differently depending on the context, but most often maya is used to personify the cause of ignorance. Figuratively speaking, maya is the trickster, the grand illusionist, the one that takes advantage of our blind spots and sets a trap for us to fall into.

Maya is a little like getting your world view from an unscrupulous cable news network. As viewers of the news network, we are kept ignorant of serious issues and the real business of government officials, not because we are unable to understand the issues, but because of the network's strategy of concealment and projection. The unprincipled news network hides the truth from us and in turn, replaces it with its own convenient, alternative reality. The presentation is so convincing and seems so real that we never question the veracity of its reporting. At some point, not only are we under its spell, we are inspired to take action against the "injustices" we perceive. These qualities of concealment and projection are how maya does its work on each of us, sometimes with tragic consequences.

In addition, maya is also responsible for persuading us to believe that certain objects will provide us with lasting pleasure and happiness. Maya does this by hiding the negative aspects of an object (concealment) and emphasizing the beneficial (projection). We know that no object can give us permanent happiness and yet we continue to chase them ignoring the simple fact that they are impermanent, inert and incapable of emanating joy. In short, we choose to believe in object-oriented happiness in spite of all contrary evidence and continue to do so mostly, because everyone else does.

The apparent expiration date on an object's joy-emitting powers occurs because, like everything else in the samsara, our body and mind are constantly changing and with them, our preferences. Familiarity breeds boredom and because we're always looking for something more, better, and different— something that will give us those initial few seconds of excitement once again—our search for the next pleasure-inducing object never ends. Maya is what keeps us on a treadmill, putting the carrot of unfulfilled desire just out of reach so that we eventually end up on our knees from sheer exhaustion.

Another of maya's seductive cons is its power of subject-object reversal. While most people don't have a problem discriminating between themselves and objects "out there," where many of us struggle is discriminating between ourselves and the objects that appear within—namely, our thoughts, feelings and emotions. Throughout the day we identify with our internal condition by telling ourselves "Now, I am happy," "Now, I am bored," "Now, I am upset," "Now, I am happy again," and so on. Or we berate ourselves for having bizarre thoughts that make us feel shameful and challenge our sense of integrity.

Worse, are those thoughts and feelings that arrive in the form of a strong desire or fear because they have the potential to influence our behavior and actions in negative ways. We all have experienced being driven by uncontrollable desire or fear. This sort of manic behavior is usually what gets us in trouble and is almost always accompanied by suffering.

But all is not lost if we simply stop and ask ourselves, *how can I be that which I know?* If I'm the knower of my thoughts, my feelings and emotions, logic indicates I must not be them. And indeed, I'm not! As we'll see later, there are many false layers to our self-identity. Identifying with one's feelings is just another one of maya's many subject-object reversals.

Nothing Is What It Seems

Seekers are often confused when they hear the wise suggest that life is a dream. But basic high school science shows us that all objects are ephemeral, fleeting and without substance. Even one of matter's most basic constituents, the atom, is mostly made up of empty space.

When you push against a wall you aren't in actual contact with anything solid. Instead, what you feel are electrons from the atoms in your fingers pressing against the electrons in the wall. The reason for the feeling of repulsion is because it takes energy to push two atoms close together and re-pattern their electron "dance" — which is more energy than your muscles can provide. When examined closely, objects are really just clouds of tiny particles held together by electrical forces. What you are touching are just repulsive energy fields. So, any feeling of solidity is just an illusion.

But the magic doesn't stop there. In many ways big and small, nature shows us that our mind is doing some amazing things to make our experience literally "come to life." Take light, for example. Light is just made up of invisible magnetic and electrical fields (more energy fields!). We know that neither magnetic or electrical fields have any visual properties and yet, somehow our neural circuitry creates colors and patterns out of them. It's all just stimuli that our brain uses to produce an image in our mind so that what you see *isn't* what you get. From this perspective, life may be seen as an elaborate stage constructed to work correlatively with the senses.

The mind's role is that it acts like an undetected function in the background constructing our apparent reality. The same life force that digests our food and circulates our blood, is the same life force that creates the world in consciousness. Like our bodily functions, the mind works so seamlessly to create our experience that we aren't even aware it's happening. Our error is that we take the world to be solid, unchanging and real.

Because of this dream-like quality of our experience, another fundamental truth we find in samsara includes:

The belief that objects are substantial, unchanging, independent and always present.

Although we interact with objects every day, because they are actually insubstantial, changing, dependent and not always present, they are as good as not real. They are like a mirage. If our experience of time were faster, it would be obvious that objects are nothing but temporary appearances, like clouds. When we observe clouds, we easily accept the fact that they won't appear the same or appear at all within even a few minutes. Unfortunately, it's difficult to observe the same truth when observing other more concrete objects such as a car, house or even a mountain. In the end, these objects too, are only temporary name and form. However, we do experience objects so we can't just write them off completely. Therefore, we say objects are *apparently* real, meaning that objects are experienced but lack intrinsic substance.

Vedanta, an ancient wisdom tradition derived from the Upanishads, comes to the same conclusion as the scientists. While Vedanta is sometimes described as a science, its objective isn't to show how nature works. Instead, Vedanta uses experience to empirically arrive at the principle that existence is non-dual—in other words, that all objects resolve into awareness. To make the point, Vedanta traditionally teaches the location of objects, as well as the three states of consciousness: wake, deep sleep, and dream states.

The teaching on the location of objects works very similar to what science already tells us. It begins with stimuli, which Vedanta calls a power or maya. According to Vedanta, maya doesn't exist "out there" because that would make existence dual instead of non-dual. Vedanta says maya, the power that creates the world, occurs in awareness but isn't a part of awareness. This is why maya is so mysterious and just a tad baffling—it's neither a part of awareness nor outside of it. According to Vedanta, these stimuli occur when we make contact with any of the five basic elements (space, air, fire, water, earth) via the sense instruments (eyes, ears, skin, tongue, nose). The sense instruments are simply data inputs that send data (the properties of the object) to the internal sense organs (the mind).

In spite of what we might believe, this data doesn't come in registering as "table," for example. Instead, the object is perceived by the internal sense organs as individual properties such as *brown, hard, smooth,* etc. No sense organ perceives an actual substance or object. For each property there is a corresponding sense organ so that, for example, the sense organ that correlates with the eyes only detects the properties of color and shape, and the sense organ that correlates with touch only perceives the properties of pressure, temperature and texture.

The sense organs channel the aggregated information to

the mind which—without any effort on our part—traffics the incoming information and applies a name to the form so that *wet, cool* and *blue* become "water," and *hot, yellow,* and *flickering* become "fire." If you follow the logic, you'll understand that what we perceive as "fire" is just a thought and its location isn't "out there" but in the mind. In other words, we don't live in an object universe, we live in a thought universe! We never have actual contact with any objects because everything is constructed in the mind. Once again, it's shown that all objects are just stimuli and our reality, mind assembled.

Now, if this isn't all obvious to you don't blame yourself. As previously mentioned, ignorance is hard-wired. We are programmed to see the world as "out there," not located in our mind. It's as if we were all born with one of those virtual reality players strapped to our heads and our parents somehow forgot to take it off. We assume the VR projection is real because we don't know any better. The appearance is so real, so lifelike that it fools everyone. In short, that's maya. Maya is what makes the impossible, possible.

Maya is what makes us not realize, for example, a pot is just clay or a shirt, just cotton. The shirt exists, but it's not really there, it's just a construct. So, with Vedanta, we're able to say that an object can both exist and not be real at the same time. And as basic science confirms, just because something appears to be one thing, doesn't necessarily mean it's so.

Vedanta also uses the three states of human existence— waking, dreaming and deep sleep—to make the same point that objects are not real. It starts by stating the obvious: what you call a dream is just you. Your dream is like a movie that is simultaneously directed, produced, acted and watched by you. In the dream state you are both the subject and the objects. You are the witness, the house, the people, the weather, the trees... the whole creation! Where else would your dream come from? Which brings us to the question, how is a dream different from the world you experience when awake? A dream feels real while you're in it, it's only when you awaken that you realize it was all a dream. Just like the teaching on the location of objects, the dream-objects are just a thought located in your mind.

To summarize, Vedanta makes several points to show that objects are only apparently real:

- Objects are temporary, fleeting, impermanent, uncertain, unstable, limited, and changing from one moment to the next. All objects have a beginning and an end.
- Objects are dependent on other factors for their existence; they are made of parts. What constitutes a cat? If the cat loses a leg is it still a cat? What if it loses everything we assume makes a cat: legs, tail, whiskers, fur, ears, and fangs. Is it still a cat? How do you define the essence of a cat? All objects are just an aggregate of other objects, a temporary arrangement of parts governed by universal laws.
- Objects are divisible. Shirt —> fabric —> thread —> cotton —> fiber —> cellulose —> carbohydrate —> molecules —> atoms —> electrons, protons and neutrons —> quarks —> all the way to awareness. Vedanta says awareness is the smallest derivative because without awareness, nothing exists (just like you can't have a movie without a screen).
- Objects are only a play of inert and insentient elements (the Five Basic Elements as defined by Vedanta are space, air, fire, water, and earth).
- Objects cannot be verified, because what you're experiencing is actually only the properties of an object (color, shape, texture, weight, smell, sound, taste) which belong to the senses and not the objects themselves.
- Objects don't exist in all states of consciousness. Where is the world when you're in deep sleep? It is unknown, as are all objects.

So, now that we've explained why objects are not real, what are they?

Objects are:

- Gross and subtle forms. For example, thoughts, feelings and emotions are all considered subtle objects.
- Inert; they are the basic elements as described previously.
- Maya—That which makes the impossible, possible. Maya is a mysterious power, also referred to as the causal body, or the unmanifest seed form from which all objects come, as well as the *gunas*. The gunas are the essential constituents of all forms and include *sattva* (knowledge), *rajas* (energy) and *tamas* (matter). The gunas are also found in subtle forms such as thoughts. The psychological aspect of the gunas is often translated as sattva (clarity, beauty, peace), rajas (desire, passion, projection) and tamas (sloth, dullness, concealment).
- Thoughts. As we learned from the teaching on the location of objects, all objects are just thoughts located in awareness. Vedanta asks, "How far are the objects from your mind?"
- Name and form—A construct created by the mind, derived from properties the sense instruments detect and the sense organs perceive. When shown a piece of jewelry, we might say it's a gold bracelet but actually we should say it's bracelet-y gold. This apparent reversal is the power of maya.
- *Mithya*—All objects are superimposed over a substrate (awareness). The definition of the Sanskrit word mithya is "apparently real" because that which is mithya is as good as non-existent. The sages like to say the world is like a dream because nothing lasts, nor is it what it appears to be—it's mithya.
- Awareness. Ultimately, all objects resolve into awareness.

The objects are me, but I am not the objects. What this seemingly cryptic phrase means is that all objects are constructed out of awareness, but they aren't who I am. Who or what I am *is* awareness—the screen upon which all the objects appear. Vedanta says the existence of an object belongs to awareness, not to the object itself. Maya reverses the relationship between existence and an object so that, for example, a mountain appears to exist. But the truth isn't that a mountain exists, but that existence "mountains." Just like from clay a potter is able to shape an infinite variety of pots, cups, plates, etc., maya adds name and form to awareness to produce an infinite array of objects. Another way to look at is by seeing awareness as being similar to water—it has no particular form, but it can assume any form.

At this point, you might be asking yourself the reason for wanting such object-related knowledge. What's the utility of knowing samsara isn't real? In fact, it might feel a little unsettling knowing that nothing here is actually real. If objects aren't real, then like the sages suggest, life is just a dream existence, right? But we needn't take a nihilistic view. Once we have the knowledge that objects aren't real, we can begin to realize some of the benefits:

Happiness isn't in the object. If objects aren't real for all the reasons mentioned above, then they must not hold the happiness I believe them to have. That means the happiness I believe objects contain is really just me, awareness.

I am that which is beyond all objects, including my body. Objects are as good as non-existent because they have no effect on me. If I am awareness, then ultimately, I have nothing to worry about because nothing can touch me. I am immutable.

This knowledge is different from other kinds of knowledge such as physics, chemistry or biology. With science the close examination of objects only reveals more ignorance (that there's more to know), but Self-knowledge is that when known, everything else becomes known.

Chapter 3

Layers Deep

Once we start to look at all the ways ignorance, or maya, fools us, we might begin to wonder just how far this ignorance goes. Maya's ability to conceal the truth is endless, but its biggest fake-out is how it makes each of us believe we are a breathing, walking, talking, thinking, eating, drinking, object-pursuing, ambition-chasing, pleasure-seeking person (the enjoyer/doer). It's one thing to trick someone into believing they have a rabbit in a hat and quite another to trick them into believing they are a walking, talking person. So, what's happening?

What's the trick?

Obviously, we are a person with a name, history and preferences—a necessary social convention that helps us navigate the complexities of the world. And in order for this world experience to work, we need to appear as many separate individuals, each with his or her particular role and set of tasks. However, simple self-inquiry reveals there is no "me" to be found anywhere and that our ignorance goes many layers deep.

Again, just like with thoughts, the rule of thumb is *you can't be that which is known.* You can't be both subject and object for the same reason you can't see your eyes with your eyes or taste your tongue with your tongue. This simple observation rules out all kinds of erroneous beliefs we have about ourselves and at the same time, helps cut through maya's deceit.

Let's start with the body, the outermost of maya's veils: You can't be the body because if you were, you would be constantly occupied managing all your bodily functions, including (but not limited to) circulating your blood, processing your food, producing your hormones, eliminating your waste, ensuring your nerves are communicating with your brain, defending off disease, and more. Furthermore, if you were your body you would have many more choices, like the option to never sleep or sleep all the time, never eat or eat all the time. You could also decide, for example, what length to grow your hair, or even to grow an extra limb.

You're obviously not the body because when you look carefully in the mirror all you see are stitched-together mom and dad parts with maybe a few here and there that skipped a generation. Perhaps you have your dad's steely blue eyes or your mom's lovely hands. In fact, there's nothing original about your body except for the compositions of its parts! Your body is nothing more than an ancestral gene map that goes back… well… forever. The body is yours to keep healthy, but other than that it's just copied parts and "you," a temporary occupant in a string of inherited biological patterns.

What about your mind? If I'm not this strong, handsome, perfectly framed body, am I not this brilliant mind? Nope. If you were your mind you would be able to say what you will be thinking in five minutes, or program it so you only have happy thoughts. How many times has the mind worked to sabotage you by saying things like, "I'm not good enough" or "I am a bad person." What kind of masochist would tell themselves these kinds of things? You are not your feelings either, for the same reason you're not your thoughts. Who would choose to be happy one moment and depressed the next as if happiness were too boring and debilitating despair the next best option?

And as far as your sharp-witted intellect goes, let's face it, it's not always that sharp! Like a knife, it can quickly become dull and needs constant honing to keep it in good shape. A sharp intellect is also dependent on how much sleep you get, what's ingested in the body and whether or not you're in good health. Where's the intellect when you're sick or with a headache? Like our memory, the intellect isn't always accessible, so it must not be me either.

That leaves us with the ego or the "I-sense"—maya's last hideout; the one place maya knows no one will find it because it's that part of us that believes it's the enjoyer/doer. Unfortunately, the ego is just a phantom, a kind of story we tell ourselves based on our history and preferences. Again, this is useful for getting around in the world, but due to our attachment to it, a hindrance to arriving at the truth.

Again, you can't be any of the things you believe you are because all these things are known by you. This means I must be that which knows. And I am! I am that which knows it is experiencing. I am that "I am"—the Self.

With each of maya's veils, there exists the illusion of continuity, separateness, control and solidity. In spite of all contrary evidence, we still believe the body-mind complex is substantial and that it will continue indefinitely as a separate entity in control of its destiny. And yet, as we've already seen, this can't be true.

Which brings us to another belief about samsara:

The belief that I am the doer and a separate entity among other entities.

Like the other entrenched beliefs that inhabit our thinking, this belief also sets us up for a host of negative psychological conditions that keep us on the hamster wheel of samsara. For example, if we identify with our body-mind complex we will suffer when we see it weaken, become sick and near death. We might also become frustrated upon realizing we aren't able to control it, make it look more beautiful, bigger, stronger, or smarter. You have to ask yourself, with so much potential for disappointment, why would you ever want to identify with it in the first place?

But maya's trickery doesn't end there. After much inquiry and discrimination, we might finally be able to convince ourselves we are not the body-mind, but that still leaves us with the belief that the world exists with many separate entities (persons) walking on the surface of the planet. In short, our ignorance doesn't end until we realize that existence doesn't equate to many, but only to one.

Like all objects in samsara, the doer is divisible. It's divisible not just from a physical aspect (the body is made up of the same basic elements found everywhere in nature) but also from a subtle aspect (thoughts, feelings, emotions, memory and intellect). The doer is divisible right down to awareness because awareness is where you arrive at by peeling away all the layers of the onion.

Awareness is that which is not divisible, made of parts or dependent on other objects. Awareness is that which enlivens these bodies and makes the world possible. We all know you can't have a living being without awareness/consciousness. Vedanta takes it an extra step by saying not only can you not have a living being, you can't have a world without awareness/consciousness.

Awareness doesn't create living beings or the world (for that you need maya) but it does make it possible. It's like the electricity required to keep all your digital devices up and running. And yet, we don't say there are *electricities*, just electricity. And like electricity, awareness doesn't come in different shapes, sizes, weights and colors—it's one size fits all. The bottom line is there is only one awareness and you are it! Awareness is the essence of you. In the end, you are a "what" not a "who." You're not even the knower of your thoughts, you are that which illumines the knower, the knowing and the known.

The goal of the seeker isn't so much to gain this knowledge as it is to eliminate ignorance, because you already are that which you seek. This is why Self-inquiry is often referred to as "the pathless path," because there is no path if, upon careful observation, you find you are the destination you seek.[1] The reason why this simple fact eludes us is because awareness isn't an object, it's the subject. Looking for awareness is like looking for the camera in the photograph—it can't be found, only inferred. We know the Self (awareness) is there, but we're unable to detect it with any of the senses.

While samsara with its many layers, twists and turns may seem to take us down an inescapable rabbit hole infinitesimally deep and complex, there is a way out. Like a nicotine habit that won't go away with more smoking, samsara has no end, but it does have a way out and understanding the essence of what you are (pure, infinite, non-dual awareness) is the key.

Endnotes

1 Once while traveling on a highway in Ecuador over the Andes, we drove past a cement guardrail that was preventing cars from going over a very steep cliff. On that barrier was graffitied in large letters, *Dónde vamos que nunca llegamos?* which roughly translates, "Where are we going such that we never arrive?" That question stuck with me for a very long time. Little did I know that it was pointing to that which is I was seeking and would find some 25 years later.

Chapter 4

The Game of Samsara

Now that we've peeled back the onion of ignorance and have seen how far it goes, we can begin to look at some of the other features of samsara. Samsara is like a game; in this game the winners are those who are able to discriminate between what's real and what's not. The rest are those who take samsara to be real and repeat the same mistakes over and over expecting a different result (the definition of insanity). The game eventually exhausts the latter group and forces them to look at what they're doing wrong, make changes, or continue to experience the same disappointment and pain once again. The game's playing field is called "the Field of Experience"—that which provides the opportunities to learn, adapt and eventually, "exit" the game via liberation (*moksha*).

From the previous chapter, we already know that what is not real Vedanta calls mithya and that this refers to all objects, including the body-mind. That only leaves us with awareness, which is not an object because it is the subject, *satya*—that which remains the same and is always present. Even in deep sleep, awareness is still there. Otherwise, we wouldn't know to wake up when the up-stairs neighbor comes home late and turns on their music.

The key to understanding both mithya and satya is the knowledge that they co-exist but in different orders of reality. What effects mithya (maya, the material world, our empirical experience), doesn't affect satya (pure awareness). Mithya is the moving images on the immutable screen of satya. The moving images projected on the screen might show world-collapsing destruction, but the immutable screen won't even reveal a blemish. As individuals, we are each playing a different movie, but the screen is one. My movie won't be the same as yours, but the screen is (and can only be) one and the same. In the end, I am/you are/we are unchanging non-dual awareness, the one without a second.

The word *karma* in Sanskrit, signifies action or deed and its cause and effect. Karma is the momentum of our past actions. In the Field there are the players who, due to ignorance, perpetuate a never-ending cycle of becoming by way of their desires and actions. The Field is regulated by laws and merit so that accruing good merit via appropriate actions ensures one a good future existence and accruing bad merit via inappropriate actions ensures a bad one. When we are familiar with and follow the rules of the Field properly, things tend to work out for the better because we are in harmony with the rules and how reality functions. When we don't, we may find ourselves stuck in samsara.

The players, or *jivas*, play out their lives and are characterized by their identification with objects (including experiences and relationships) and their desire to reap the fruits of their actions. Each player is an apparent living creature created and maintained by maya and made up of three bodies vivified by awareness. The three bodies are the Gross body (physical body), the subtle body (mind/intellect/ego), and the causal body (the jiva's innate programming or sub-conscious). There are other kinds of players in the Field too, but their subtle body is less developed, and they don't experience life in the same way humans do—that is, they are not self-aware. These other jivas include animals, insects, plants and microbes. Like the human players, they are also a product of maya vivified by awareness, but lack the intellect bestowed to us humans.

A prominent characteristic of the Field is its pair of opposites. The Field is a strange place but if one understands how it works, one feels less impacted by its changing nature. The human jivas love to lament about the Field along with all its limitations and perceived injustices, but the problem isn't the Field, it's the jivas. As the wise like to say, "you are both the problem and the solution." The reason for this is simple: all problems exist in the mind. We suffer not because the world is imperfect but because our understanding of it is. We can't change the world, but we can change how we see it in a way that better aligns with reality. In fact, it's only when our thoughts don't align with reality that we suffer.

The first thing to know about the pairs of opposites is that the Field is a mixture of pleasure and pain—you can't have one without the other. Life can never be all pleasure, nor can it be all pain. Even the worst circumstances have a little pleasure and even the best experiences, some pain. You can accept both pleasure and pain or refuse pleasure and pain, but you can't take pleasure without pain. Even those who are enlightened experience some pain but are less affected by it due to their understanding of the law of opposites and their belief that forbearance is a virtue. Instead of emotionally reacting to pain and thereby increasing its sting, the wise intelligently respond to it, letting it roll off them like water to a duck. They know pain is a part of life, is only apparently real (mithya), and don't identify with it. And as far as pleasure goes, the wise also recognize that it's just mithya—to be enjoyed but never to be attached to in the form of longing or dependency.

Related to pleasure and pain is the fact that life is a zero-sum—for every up there's a down and for every down there's an up. Just look at the lives of the rich and famous: In spite of their good looks, talent, money and numerous connections, they still can't beat the karma system. Why? Because they are governed by the same physical, psychological and moral limitations that all mortals are. The biggest reason they suffer is because they normalize their success and begin to identify with it, which causes suffering when it changes and is no longer present. Like us, they also like to challenge the game rules and see what they can get away with—they just do it in more style and comfort! And while we foolishly worry every day about not having enough, they worry every day about keeping what they already have—including their good looks, talent, money and connections. Where's the gain?

Of course, the pairs of opposites extend much further than just pleasure and pain. The Buddhists speak of the "eight worldly conditions"—pleasure and pain, loss and gain, praise and blame, and recognition and insignificance. It's the nature of the Field that these pairs of opposites are constantly changing. To insist that they should never change is to negate the natural course of things. In other words, it's only when we count on pleasure, gain, praise and recognition and not their opposites that we set ourselves up for disappointment. In the Bhagavad Gita, Krishna suggests Arjuna have an attitude of sameness as he goes into battle to defend dharma. This isn't indifference, it's the knowledge that, although we may choose our actions, we are not in charge of the results. The wise practice equanimity without attachment or aversion, seeing life's events as nothing more than the changing worldly conditions.

Actually, the entire Field is dependent on pairs of opposites for its ability to function. You can't have sweet without sour, dry without wet, or hot without cold. Part of what makes life enjoyable is its variety and contrasts. How would we know beautiful without ugly? Fun without mundane? Happy without sad? If everything were beautiful, we would have nothing to compare it with. The fact that we don't experience beauty every moment of the day means that when we do, we are pleasantly surprised. Thus, maybe we should appreciate the mundane, because without it, beauty wouldn't exist! Without the pairs of opposites there is no game, which brings us to the next topic— the rules of the game.

In order for any game to work, there must be a set of rules. Animals, insects, plants, microbial and other beings already follow the rules to a T. The same can't be said for humans. Animals are programmed to follow the rules and not question their experience or wish the rules to be any different. Their minds are perfectly hard-wired to accept their innate nature and follow through with their tasks, whether it be gathering acorns or hunting wildebeests. It's only humans, with their endowed intellects, that get in trouble devising clever ways to subvert the Field (usually in an effort to squeeze a few more drops of pleasure from it). Only humans attempt to create their own reality through their use of beliefs, stories, fantasies and lies. Only humans willingly create an environment of addiction coming up with new and better ways to make it easier to satisfy their endless desires. Only humans seek out help from gurus, priests, and psychiatrists, because only humans are confused about their purpose and place in the world.

Samsara has its rules and it's only when we know and follow them that our worries may lessen. For example, there are universal physical laws that keep us from damaging the hardware (these bodies). The physical laws governing what we can and cannot do should be obvious to anyone who has ever stuck their finger in fire, tried to eat dirt, decided to never sleep again, or attempted flight from a high perch. Without physical laws, human beings probably wouldn't make it past the age of two! The body brilliantly protects itself by immediately alerting us to its limitations. Unfortunately, there are those who try to thwart the system and either end up at Emergency or dramatically shortening their life expectancy. Physical laws also limit the amount of pleasure we can be exposed to. As sentient beings, we can only eat so much, drink so much, sleep so much and fornicate so much before nature converts it into the opposite of pleasure—pain.

Next, are universal psychological laws. Because of our intellect, these laws mostly apply only to us. We all know that humans can become damaged both physically and mentally due to certain behaviors and circumstances. Violence is an obvious way in which we are able to inflict not only physical wounds, but psychological wounds onto others and ourselves. Unless you are a sadist, abusing others probably isn't going to feel good. In this way, universal psychological laws protect the psyche from prolonged exposure to environments that endanger its health and that of others (such as with an abusive relationship or with war). Thus, the Field not only brilliantly protects the hardware but the software too.

Lastly, are moral laws. Moral laws protect us from devising ways to harm others for our own gain. Universal moral laws protect the total. Another way to describe moral laws is empathy. Without empathy we're nothing more than a bunch of rabble-rousing Vikings raping and pillaging. The effect of going against moral laws usually arrives in the form of guilt, shame or anxiety (not to mention retribution from the victims of our moral lapses). But virtue is not just for the pious. In moments of clarity, we all intuitively feel a connection with others and know our essence to be the same. While it helps to be aware of moral laws, they generally don't need to be taught because they are common sense and are a natural acknowledgement of our non-dual nature.

It might be better to think of all universal laws—whether physical, psychological or moral—not as rules, but as limitations governed by the Field of Experience. Nobody makes the laws, they just are—like gravity. For the same reason we avoid kicking big rocks, we choose to not steal from our neighbor or act violently toward others—because it hurts!

You can either accept or deny the rules but denying them in the long run isn't going to get you very far in the game. Every day we make choices and choose to play by the rules or not. Once we decide to observe the rules, we can stop kicking big rocks and enjoy a greater peace of mind (and straighter toes). However, if we continue to flaunt the rules the result is inevitably an agitated mind that lacks the clarity to see the truth and prevent suffering.

Related to the topic of universal laws is dharma. Dharma has several meanings but is often used to describe the natural order of things, as well as the timely and appropriate carrying out of actions. All things in the Field have a natural order. For example, it is the dharma of flowers to bloom, fish to swim and birds to fly, just as it is the dharma of bees to sting, snakes to slither, and lions to hunt. Everything, including humans, has a role to play in the Field and contributes to the system in one way or another in order to support the entirety of creation. Even the blood-thirsty lion with its violent means exists to help maintain the balance on some order. When examined closely, we see that nature with all its connections and laws makes no mistakes. The brilliance of the system is that everything fits together in a symbiotic relationship.

Because humans are also assigned a role in the game, dharma can be understood as responding appropriately to one's duties, responsibilities and conduct. Appropriate action results in spiritual wealth and a general feeling of well-being, while inappropriate action results in spiritual poverty and a feeling that one's soul has been diminished. One is in union with the natural order of things, the other is not. When I properly take care of myself, those closest to me and my environment, I'm acting in harmony with universal dharma and will reap the benefits. When I neglect my health, act in selfish ways and ignore my environment, I'm acting against dharma and will inevitably feel that something is off the mark.

Another way to work against the natural order of things is by choosing to ignore one's personal dharma. You can think of personal dharma as your innate programming. Just like it's the nature of squirrels to collect acorns or beavers to build dams, it's the nature of some humans to make money and others to develop urban spaces. Part of leading a harmonious life is not negating your personal dharma. We can only follow our own nature. Each time we negate our personal dharma or try to replicate that of another we're doing ourselves a disservice by setting the conditions for a mind that is constantly disturbed.

Does this mean we should drop our day job and pursue our dream of becoming a full-time poet? Not necessarily, but it does mean we should make time for whatever it is that drives us. Sometimes following one's dharma means compromising and using our innate skills for other more lucrative tasks, if that's what is required of us. Neither is following our personal dharma an excuse for flaunting our responsibilities or just doing whatever feels good. As with everything related to dharma, we need to live intelligently and choose the most appropriate action—that which is in harmony with both our personal nature and universal laws.

Lastly, is the concept of karma. As described before, karma is action and the cause and effect of action. Its role in the Field is to provide the result of one's actions. Karma is important to regulating and enforcing the rules of the game. It can be likened to growing fruit, and the Field likened to an agricultural field. To grow fruit, the human jivas sow seeds in the form of action. While the farmer is in control of the seeding and cultivating, he or she is not in charge of the results. Depending on the farmer's actions, her seeding and cultivation will either result in sweet fruit (desired results), bitter fruit (undesired results) or mixed fruit (neither desired or undesired results). When our actions are appropriate, we sow the seeds that produce sweet fruit. When our actions are inappropriate, we sow the seeds that produce bitter ones. Thus, the function of the Field is to not only provide the opportunity for action, but the results of those actions. As doers, we are compelled to sow seeds, and as enjoyers we are compelled to reap favorable results. So, if you're a good farmer you will strive to make your life rewarding by only planting those seeds which have shown to give the sweetest fruit. As simple as this sounds, there are those farmers who choose to ignore the basic rules of the game and instead look for shortcuts and work-arounds ignoring the symbiotic nature between the farmer and the Field. As a result, they are caught holding a bag of rotten goods.

The Simile of the *Ashvatta* Tree

Another way to show how samsara operates and dictates our experience is through the more traditional simile of the *ashvatta* tree. The ashvatta tree is an unusual species of fig that grows throughout India. What makes the ashvatta tree so extraordinary is the appearance of having its roots in the air and its branches in the ground. The ashvatta tree is not only a metaphor for samsara but a sort of Vedic map describing how the ancient seers perceived the manifestation of existence. The following is adapted from my book, *The Wisdom Teachings of the Bhagavad Gita.*

The highest "roots" of the inverted ashvatta tree represent pure awareness (the basis of all creation) and are described as being above in the sky because they are the most subtle. In contrast, its "branches" are below in the ground within the substructure of time and space. Because of its cyclical nature, the ashvatta tree has no beginning and is eternal. Even after being cut down, it still exists separately in its seed form.

The subterranean branches of the ashvatta tree are ranked according to height so that the highest branches represent the celestial beings, the middle branches represent the human beings, and the lower branches represent the animal and plant beings. Sense objects are its buds. By jivas desiring sense objects, new beings are born in the form of more branches. So, the ashvatta tree is a tree of becoming.

A jiva is said to be reborn based on the kind of karma they accrued in their previous life. If their account shows mostly good karma, they might be reborn in the higher celestial branches.

If, instead, they had accrued mostly *bad* karma, they might be reborn in the lower animal branches where they are given the opportunity to exhaust their *adharmic* tendencies. However, no level of branches is ideal, for once the jivas in the higher branches use up all their good karma, they are sent back down to the middle branches (the branches of humans jivas) where they are provided once again, a chance to gain good karma. In this way, the jivas are said to spend endless lives chasing objects and coming and going within samsara.

One is said to cut down the ashvatta tree with the strong axe of detachment. According to scripture, the tree is firmly situated in the ground due to ignorance where it can only be uprooted by Self-knowledge. Only time can provide enough painful experiences for the jiva to develop the desire for knowledge. This desire to know is what gives rise to inquiry, eventually leading to the knowledge that uproots ignorance.

So, in spite of maya's powers of concealment and projection, there is a way out but it involves work. To cut down the tree of samsara requires the strong axe of commitment because it's only with a strong desire to be free of samsara that one will gain the desire for knowledge. In addition, like chopping down any tree, a single swing of the axe won't suffice. It requires effort, along with much repetition. How much effort? As many swings of the axe as it takes to fell the tree.

To summarize the teaching of the ashvatta tree, we can draw comparisons between it and samsara by outlining their common features:

Both are expansive — Like the sub-roots of the ashvatta tree, samsara's bondage is spread widely due to our relationships with others and our attachments. Each of us consciously and unconsciously promotes ignorance of reality through our speech and actions. In modern times, the roots of this collective bondage extend more broadly via the internet, the media, advertising and other forms of modern communication.

Both are beginningless — A tree may have such a network of roots and offshoots that its origin is no longer recognizable. If you've ever grown blackberries (a tenacious and prolific plant) you will know that every part of the plant above and below ground is able to take root in such a way that it's difficult to tell where the blackberry plant begins and where it ends. This makes it easy to propagate the blackberry plant from both root and leafy stem cuttings, further blurring the lines between what constitutes a plant. In the same way, samsara/maya/ignorance is beginningless. It comes with being human, perpetuating itself via karmas (actions).

Both are inexplicable — Which came first, the tree or the seed? Or in the case of samsara — the desire or the action? Desire causes action, and action causes desire. This is why no matter how much we try to make ourselves feel complete via worldly means, we can never feel satisfied — because there is no resolution. It's just desire-action-desire-action, ad infinitum. According to karma theory, this same propensity to be locked into an endless loop of desire-action is the cause for rebirth and continued suffering. This is also why it is said there is no end to samsara, there's only getting out.

Both have roots — Like a tree's taproot which cannot be seen but can be inferred, samsara has its taproot in awareness — which cannot be seen, but can be known.

Both have branches — The branches of samsara are described as being made of the basic elements and being nourished by maya. They are also organized by higher branches (celestial beings), middle branches (human beings) and lower branches (animals and plants).

Both produce fruit — The fruits of samsara are the fruits of action, the results of both good and bad *karma*. Thus, there is sweet fruit (desired results), bitter fruit (undesired results) and mixed fruit (neither desired or undesired results).

Both have inhabitants—Where one finds fruit, one finds birds squawking and competing for the sweetest fruit. In the same way, in the tree of samsara there are beings perched on the high, middle and lower branches making much noise and fuss each time they experience elation or pain from the gain or loss of pleasure.

Both sway in the wind—A tree sways in the wind, while human beings sway from the push and pull of karma (cause and effect). Beings are taken here and there within their worldly experience as if propelled by an invisible force. Beings move from one birth to the next based on the sweet and bitter fruits of their actions.

In spite of what society would have us believe, life's winners are not those with the most material objects and wealth. If that were the case, the rich would literally build themselves a golden escalator to heaven. In order to be successful in the game of life, we need to understand its nature and discriminate between what's real and what's ephemeral. Only then can we enjoy limitless freedom as unchanging, eternal, non-dual awareness.

Chapter 5

Impersonal Forces

Vedanta has an elegant way of explaining the basic qualities that make up our outer and inner worlds. These powers are the impersonal forces imbued in everything. They provide the knowledge, energy and matter to make all of creation possible. *Triguna yoga*, or the practice of managing the natural powers found within, is a way to monitor overall wellness and address daily internal issues that can run amok in samsara.

In general, the gunas (meaning quality, or "rope" due to their potential to bind) are a simple and practical way of explaining the forces of nature. The same forces that create the apparent outer world, including our body, are the same forces that create the apparent inner one, including our thoughts. Ultimately, these powers are what nature uses to create, sustain and destroy/recycle all objects, gross and subtle.

These three powers consist of sattva (knowledge), rajas (energy) and tamas (matter). To grow a great towering oak, you first need sattva—the "program" encapsulated in an acorn that knows how to grow it. Without this intelligence, there would be no growth (and no acorn). We may not be able to see this knowledge using our senses, but we can infer that it exists. Of course, for that acorn to grow into a tree requires energy. So, rajas provides the vitality for growth, movement and sustainability. Lastly, we have tamas which is responsible for the actual material or matter that makes an oak. Tamas represents the inert physical contents of the oak and that of the universe. So, one way to summarize the contents of maya's world is by understanding it to be knowledge-energy-matter. Everywhere we look, we see knowledge-energy-matter working together in various combinations to construct the objects we experience.

Just like seeing an oak in an acorn, our own inner condition is a microcosm of our outer one. In other words, the condition of the mind is the same gunas expressing themselves, but inwardly. Throughout our day, we might experience a variety of feelings including clarity, anxiety or dullness as the gunas color the mind and as one takes prominence over the other. If sattva is dominant, it will bring feelings of: lucidness, peace or happiness; rajas—passion, desire or restlessness; and tamas—dullness, lethargy, or ignorance. These powers come and go through us moment by moment like the changing weather.

Rajas helps kick us out of bed in the morning and bring home the bacon but is also associated with maya's quality of projection. A mind with too much rajas lusts for objects and exaggerates their qualities. Not only does too much rajas create anxiety and aggression, it diminishes our ability to discriminate and make wise choices. People who are too rajasic act first and think later. They tend to be reactive and quick to project their subconscious content onto the world. Rajas can be compared to a windy day that distracts with its howls and stirring of dust and leaves. It can be difficult to see through such distraction and turbulence.

Tamas also deludes its host but instead, with its ability to conceal. Sometimes we welcome tamas—for example, when we want to relax with a glass of wine or fall asleep. Where tamas becomes problematic is when it hides the truth from us. Tamas contributes to attachment of desirable objects by concealing their negative aspects. People whose temperament is tamasic are unmotivated and always looking for the easy way out. Feelings of fear and doubt are predominant with them. Tamas is like a dense fog that blocks the light and dulls the intellect.

Sattva is the opposite of whirl-wind rajas and foggy tamas. Sattva is clear, sunny skies. When we are sattvic we can see the path and obstacles in front of us, thereby avoiding the traps of samsara. People who are sattvic enjoy the use of their intellect, pursue the truth, generally have an appreciation for education and the arts, and seek balance. However, sattva itself can become binding if we find ourselves clinging to the pleasure we gain from it. The starving artist is an example of someone who is deeply attached to sattva, willing to sacrifice his or her health in the name of their art (or more specifically, the feeling they get from doing their art). Spiritual types can also easily become enamored with sattva from the bliss experienced by doing various practices.

Clinging to anything is not healthy, not even the desire for clarity or peace, which like all experiences in samsara, are of the nature to change. Nevertheless, sattva (without attachment to it) is still what one should try to cultivate and maintain if they're interested in getting closer to the truth.

From a psychological perspective, total rajas might be thought of as mania, and total tamas as depression. Someone suffering from bipolar disorder has lots of rajas and tamas operating with little sattva. Like bipolar disorder, these two gunas work together. When the host is exhausted by too much rajas, tamas kicks in or may be induced in the form of a numbing device like binge-watching, alcohol, or drugs. Much of the world powers up with caffeine (rajas) in the morning and powers down with booze (tamas) at night. Many of us are prescribed anti-anxiety medication (tamas) because we can't handle the stress (rajas) of everyday life. It's a sign of the times that cannabis (tamas) has now been made legal and has proliferated in many parts of the United States. But it's not just in the U.S. where too much rajas is an ever-present problem. Extreme rajas driven by technology and rabid consumerism is like a virus that has spread to all parts of the globe endangering the Earth's balance.

The problem isn't that the gunas exist but that we are unable to manage them. Again, just like an oak tree, we need all three in various degrees in order to operate in the world. Most people are unconscious of these impersonal forces. They assume the roller coaster of life, with a predominance of rajas and tamas, is just the way it is. However, there are many things we can do to manage the gunas. First and foremost, is to not identify with them and to see them as the separate, natural phenomena they are. In order to do this, we need to be mindful of our thoughts and emotional states. Once we're able to observe our inner state with objectivity, we can seek a balance, changing our condition by avoiding certain thoughts and circumstances, while encouraging others.

Guna management is greatly affected by what we take in via our senses and can be summarized by the truism: *You are what you eat.* Everything we consume affects us in ways beneficial, neutral or harmful. This includes thoughts, speech, relationships, work, entertainment, recreation, food, medication, sex and more.

Guna management always starts with what you feed your mouth, eyes, and ears. Viewed from this perspective, junk food isn't just a bag of greasy chips, it's also all the other cheap and nutritionless content we ingest through our senses on a daily basis. You know it when you've had too much because rajas or tamas begin to kick up in unpleasant ways. So we try to avoid certain substances, relationships and activities that we know will make us feel off (too rajasic or too tamasic), and welcome those that make us feel calm, composed and at ease (sattvic).

Mindfulness of the gunas means that when we feel especially full of desire (too much rajas) or full of fear (too much tamas), we try to understand the causes and if we can't, we simply let it pass refusing to take ownership of it (like the weather!). Sometimes there is no cause to be found and the only thing to do is to let the thought, feeling or emotion run its course. Or perhaps, we do our best to counter our agitation by strategically using a little rajas in order to counter too much tamas (like going to the gym), or a little tamas to counter too much rajas (like getting a massage).

In the end, our well-being may depend on how objectively

we are able to view the world. If we're able to recognize that there is no "inner world" or "outer world"—that it's all just one world with the same natural forces at work—we can begin to come to terms with our condition and learn to work with nature.

The basic method for managing the gunas is to ask yourself how you feel after consuming or completing an action. This might involve eating, watching, listening, or thinking. Good guna management often starts with the action of eating or drinking because without monitoring what you eat and drink and how it makes the body feel, it's very difficult to monitor the more subtle gunas occurring in the mind. Most of us have a caffeine addiction that gives us an artificial boost but at the same time, obscures clarity due to its tendency to generate too many thoughts and make the mind unstable.

So, we start with diet, trying to cut out or minimize any food, drink or other substance that makes us feel overtly rajasic (e.g., caffeine, sugar, etc.) or tamasic (e.g., carbs, alcohol, cannabis, etc.). We may also observe the frequency, quantity and quality of the food we eat and adjust appropriately. People react differently to different kinds of foods, substances and amounts. Only you know which foods make you feel too much rajas (anxious, nervous, loss of composure) or too much tamas (lethargic, unmotivated, down). You might even start keeping a journal that records how certain foods, drinks, supplements or substances make you feel. By noting which substances make us feel bad, we are less likely to take them when offered the next time.

Just as important as identifying which foods or substances make you feel too rajasic or tamasic, is identifying which ones make you feel sattvic. Record and seek out those foods that make you feel clean, relaxed and healthy. In the end, it's about making intelligent choices instead of just being a slave to your conditioning. Get scientific about it and really discover what your body wants to be healthy. There is no need to hire a nutritionist or take a special physiological exam, just be honest with yourself and start recording how food and other objects you ingest make you feel.

The same management skills used to observe and control eating habits, are the same skills used for controlling everything else we consume through our senses. We've all sat through movies we wish we hadn't or have found ourselves taking in too much of the news. The same applies to books, magazines, music, websites, events, etc. Even personal relationships can be toxic and bring unwelcome amounts of rajas or tamas. Don't hesitate to keep certain people at a safe distance.

In order to manage the gunas properly, we might need to take a full inventory of our life and make a firm commitment to changing our consumption habits, or we might need to simplify, cutting out unnecessary calendar events and relationships in order to feel more sattvic, balanced and in control.

However, like everything else in samsara, it's the nature of the gunas to change. It's not possible to have constant sattva no matter how hard we try to control our inner and outer environment. In fact, our best defense for dealing with the gunas is to not identify with them in the first place. After all, the gunas are just another object on the screen of awareness, they aren't who you are. Like all subtle objects, they are just phenomena coming and going at no one's behest.

With this knowledge of the gunas, we can now begin to take a closer look at our likes and dislikes and see how they get us stuck in the mud.

Chapter 6

Stuck in the Mud

Vedanta shows us that we're pure even with all the apparent goo stuck to our cranium from years of watching TV and movies; listening to pop music and talk radio; reading magazines, books and propaganda; viewing advertisements; checking in on social media; and following the whereabouts of certain celebrities — not to mention all the other addictions we develop surrounding unhealthy foods, abusive relationships, alcohol and drugs, pain killers, sex... the list goes on and on because as human beings there is no limit to what we can become attached to. Becoming free of it all might remind us of the proverbial lotus flower emerging out of the mud. Whether or not images of pretty lotus flowers are your thing, we all get a little dirty rolling around in the mud of samsara.

For some of us, it takes years to get out of samsara for the simple reason that we are unable to convince ourselves that we've had enough of it. We keep going back for more believing that *this time* it will pay off, that *this time* I will get what I want and be totally satisfied. Most people are probably surprised to learn that objects don't have the power to make us happy. In fact, all objects are inert and value neutral. Even if joy were in the object, that would mean the same object would make everyone happy all the time. That's obviously not the case since each of us has different likes and dislikes and over time, those likes and dislikes change.

So where does joy come from?

When you desire an object, you create a sort of mental itch — a subtle or not-so-subtle anxiety. The more you think about the object, the more irritable the itch becomes until finally, by acquiring the object of your desire, you're able to scratch. The apparent happiness of getting what you want doesn't come from the object, it comes from the relief you experience from finally scratching. And once that itch is gone, it's off to the next one. For most people it becomes a perpetual rash so that their entire lives are just itch-scratch-itch-scratch. Samsara is like a mental rash. Objects don't make me itch; I make me itch. When you finally become tired of all that scratching, you're ready to get out of samsara.

The promise of fulfillment is how samsara pulls us in like a whirlpool. It starts to spin us slowly in the beginning as we enter the outer edges of its vortex. As we move closer to the axis of rotation, due to centrifugal forces, the speed increases to dizzying levels. The centrifugal force is caused by the erroneous belief that by repeating the same experience, we will eventually become satisfied. Unfortunately, what we are setting ourselves up for is a condition that, like a whirlpool, is difficult or almost impossible to remove ourselves from. Thus, we are driven by our desires and the constant reinforcement of those desires.

Vedanta calls these conditioned tendencies, *vasanas*. Every action (karma) leaves behind a trace, scent or impression. Due to our thoughts, a certain trace left behind can develop to be a strong like or dislike. With each repeated action, the momentum of this evolving condition becomes stronger, forming a self-perpetuating cycle or wheel (*samsara chakra*) so that karma leads to a vasana, which leads to desire and more karma, which just reinforces the existing vasana, and so on. This vicious merry-go-round can progress to the point where it's unclear if you're in control, or if the vasana is controlling you. This is how addictions manifest—by losing control of our ability to discriminate and do what is in our own best interests.

For example, let's say a friend takes you out to eat sushi. You've never had sushi, so it's a new experience for you. Your friend orders the "Me So Happy" combo and at first, you're impressed with the visual presentation—the way the rolls are cut, the beautiful colored sauces and the multi-layers of fresh ingredients. You pop a piece in your mouth and your senses turn on like a light bulb from all the interesting textures and irresistible mixture of sweet and salty. From that moment on, you're hooked. The memory of the visual presentation along with the texture, smells and tastes has left a lasting impression on you. A couple of days later you return to the same restaurant to order the same rolls, and sure enough, *bam!*—you have the same experience of utter delight. Now, you can't stop having thoughts about sushi. It's like a program running in your head that won't stop. You wish you could eat sushi every day!

Your sushi vasana is humming along uninhibited until one day your wife complains that you're spending all your money on sushi. You try to justify your indulgence, but she doesn't buy it and reminds you they are eating mac and cheese from a box every night, while you eat expensive sushi every day. Furthermore, you begin to have stomach problems and suspect it's from all the raw fish you've been ingesting. You become even more concerned after watching a news report about a guy who developed a 10-foot tapeworm in his intestines from eating too much sushi!

You don't know what to do. "How can I live without sushi?" you whimper quietly to yourself. Slowly, you try to make a transition and break your sushi vasana. It takes a lot of work, but eventually you swear off sushi and return to life as it was before.

The strongest vasanas often come from that which was previously out of reach. These types of vasanas are usually developed at a young age when we are most impressionable. Adults who grew up with friends who were well-off financially, but whose own family was financially limited, often have certain vasanas for material items and the status they apparently bring. In her book *Generation Wealth* Greenfield interviews a woman who spends $1000 using her modest teacher's wages to buy a Louis Vuitton handbag. When a friend finds an identical imitation of the handbag at a New York street market for $150, she wakes up to her folly and becomes dispassionate about the "piece of leather." Such are life's lessons.

Certain vasanas are hard to shake. It's a pity that so many of us must check into rehabilitation centers to try to re-program tendencies that we no longer have control over, including drug and alcohol vasanas, sex vasanas, internet and social media vasanas, eating vasanas, shopping vasanas, body image vasanas... but that's the way karma works. We learn—albeit slowly—through our suffering.

What's characteristic of all those controlled by their vasanas is that they know they are not in control. Sadly, many who suffer from addiction must hit rock bottom or worse before their behavior will change. The felt pain of continuing the habit must be exponentially greater than the pleasure derived from the object before one can begin to escape the jaws of samsara.

Just to be clear, all desire is not bad, nor is the enjoyment of objects. We needn't be ascetics and shun the world, we only need to understand its limitations. We delight in certain objects, people and experiences for the (temporary) pleasure they provide us. But we can still enjoy life without becoming attached to objects and people and losing self-control. Furthermore, it takes desire to get out of samsara—maybe, even a burning desire depending on how stuck you are and whether or not you want a little freedom or total freedom. With understanding and the right attitude, we can enjoy objects and experiences without causing ourselves unnecessary pain and suffering.

Getting out of samsara can be hard work. Many will receive multiple scrapes, bruises, and maybe even a few broken bones before they are convinced that what samsara has to offer is fool's gold. Many others will wait until a ripe old age or be on their death bed before realizing it was all just a chasing after the wind. The system is built to frustrate us by eventually exhausting and forcing us to find freedom. After all, everything we do is for freedom. For example, we might seek knowledge because we want to be free from ignorance, we might seek a relationship because we want to be free from loneliness, or we might seek money to be free from financial insecurity. Everything we do is consciously and unconsciously done in the name of freedom. And yet, samsara is not freedom, it's entrapment.

Greenfield's book is full of people who thought they could beat the system, failed, and got hurt. Samsara will suck the life and soul out of you as evident by the measures people take to hold onto money, power and recognition. We want money, but with money comes the anxiety to spend it. We want power, but with power comes responsibility. We want recognition, but with recognition comes lack of privacy. That life is a zero-sum game is no secret, we simply choose to ignore this fact because we don't want to spoil the illusion that it be otherwise. Fortunately, samsara's allure is not unavoidable and its trap, not inescapable. With even a little discrimination and discipline, we can begin to develop a dispassion for those objects that ultimately cause us to suffer.

When we desire something, we filter out reality to only emphasize the positive aspects of the object. Projecting this subjective value on an object is superimposition and it's how the mind fools us into believing we need what we desire. If we are able to also see the negative aspects of an object, we might be able to tame our desire. For example, a certain beautiful body may have great appeal until we realize it's one pin prick away from oozing itself all over the ground. Likewise, owning a Ferrari might seem like a wonderful aspiration until you imagine all the flying gravel that will be inevitably be pecking at it (not to mention jealous people with sharp keys). Discrimination means being able to tell the difference between illusion and reality. It requires an inquisitive mind that doesn't let emotion get the upper hand. When we are relentlessly truthful about what reality is presenting to us, we can see the defects in any object and maintain a sense of self-control.

What about our other two nemeses, hate and fear? Hate is just a negative desire. Maya's twin powers of concealment and projection are still working together, they are just working in a way that causes the individual to move away from the object of their attention rather than toward it.

Fear works the same way. In the previous chapter on ignorance, we used the example of the unscrupulous cable news network, which also serves as an example of how both hate and fear can develop. The reason so many of us are drawn to such news media in the first place is because, through cover-up and projection, they seem to confirm what we already hate or fear. We may have planted the seeds of hate and fear a long time ago and had only needed a little encouragement to have them grow.

Sometimes fear is justified, like the fear we might have crossing a crumbling bridge or encountering a bear on a trail, but most fear isn't. In fact, fear is more often than not, just a thought and not an actual warning of imminent danger. A fear thought appears on the screen of awareness and instead of questioning it with the use of our intellect, we identify with it and let our emotions take over. Ninety-nine percent of the time, f-e-a-r stands for False Evidence Appearing Real.

Even hatred has no justification when you realize that evil doesn't really exist, only ignorance. People are hurtful toward others because they are unaware of the cause and effect of their actions. Driven by forces and conditioning they hardly understand, they let their emotions run the show without examining their experience and considering the consequences. Empathy becomes easier when you understand that as human beings, we are mostly blind to the follies and forces that influence us. Unfortunately, due to our collective ignorance, "evil" can sometimes have a good run of it before society finally wakes up and decides it has had enough. The gears of karma may turn slowly and unseen, but make no mistake, it always delivers.

Related to hate is anger, which is the extreme agitation we experience when we aren't getting what we want. Anger is sometimes a useful and necessary emotion. It can knock us out of a state of indifference and motivate a change for good. But more often than not, anger is just another form of attachment that leads to suffering. The following shows the chain effect of both attachment and anger.

Attachment (from a binding desire or hatred) > **anger** (due to not getting what I want) > **delusion** (an inability to see clearly) > **loss of discrimination** (an inability to make a proper decision) > **loss of peace and freedom** (suffering)

The key is to not let the chain of attachment begin in the first place because once you're at the anger stage it's usually too late and you're already stuck in samsara's vortex. It's for this reason that scripture often takes a militant approach regarding the management of likes and dislikes. There's the recognition that the forces of attachment are unduly strong and require, at times, superhuman powers to overcome.

It takes practice to standup to your likes and dislikes and not allow your emotions to get the best of you. As already shown, ignorance is intelligent and can be very clever. Ignorance knows your weak points and will go to great lengths to persuade you that the perceived pleasure outweighs the actual pain. Maya is a master at making and breaking promises. This is why knowledge is paramount to freedom, because without knowledge of the magician's trick, we're played for suckers every time.

However, getting out of the mud of samsara requires a plan. We must come up with a strategy, stick with it and most of all, be patient as we begin to gradually weaken old habitual patterns. The constant reinforcement of certain tendencies has taken years to develop, so we can't expect them to disappear the moment we decide to declare war against them. We need to be mindful of any troublemaking vasanas and view them objectively.

Taking control of our likes and dislikes requires commitment, honesty and intellect. And as tough as that sounds, the rewards are a greater sense of control, confidence and a Teflon-like composure as samsara confronts us with one obstacle after another.

Chapter 7

The World Is Made of Stories

To get out of samsara we may need to challenge our core beliefs, which, for many of us, are viewed as foregone conclusions. These might include beliefs about how we see ourselves, other people and the world. While adherence to beliefs can give us a sense of empowerment, stability and comfort, they can also lock us into ideas and concepts that leave us uncertain, anxious and in the dark.

Samsara thrives on beliefs, that is, the stories we tell ourselves and tell others. Stories are the very oxygen we breathe. They are behind every decision and action, emotion and desire. They color our day-to-day experience whether it be the story about our family, community, country, or creation itself. Some of the stories we choose to live by are helpful because they have the power to motivate us toward a beneficial outcome. Others are not so helpful because they conceal and distract us from what's true, or worse, put us and others in danger. There are also those stories which confuse us because they only contain half-truths. Most religion is made up of half-truths and lacks sound reasoning.

The best advice for anyone at any age is to pick your stories wisely because similar to our likes and dislikes, stories have the power to bind. By being selective about which stories to believe in, we can save ourselves much wasted time and suffering. Some, blind to their own pursuits, might spend years following the wrong story. Often people don't realize they've been chasing the wrong story until well into their twilight years. They look back on their life with remorse, thinking it was all just a "chasing the wind."

Everywhere we go we are told stories. Not just from friends, teachers, bosses, politicians and the media, but also from corporations who spend millions to have their brand associated with a certain narrative. We might not even be aware of some of the stories we unconsciously live by. For example, the pervasive belief that obtaining power and riches is the apogee of being a successful human being. This popular social narrative is based on a bizarre conjecture that assumes riches and power bring absolute security and happiness, and that with riches and power I will be made complete and totally satisfied!

These kinds of universally accepted stories are, of course, rooted in our ignorance and feeling of lack. They, along with other popular themes, such as fame and romantic love, can keep us stuck in samsara for years, having us believe that if we just try harder, we will obtain the elusive worldly happiness we so badly seek.

Unfortunately, lasting worldly happiness doesn't exist—a simple but hidden fact that most of us overlook. Sadly, there is no real happiness to be found in samsara. Instead, what we find is our small self in conflict with a world where everything is constantly shifting and not always accessible. That isn't to say that satisfaction isn't possible. It's just that real happiness is not what you think it is and probably not where you'd expect to find it.

The stories we carry with us are not always life-affirming, either. Fear creates many unpleasant stories in society that can affect us in a myriad of negative ways. News organizations— which like to report airplanes that crash but not those that land—are constantly generating fear stories. Every day, we tune into to what amounts to mostly negative coverage of the world in order to learn about the latest screw-up, outrage and existential threat. This constant bombardment of negativity can alter the way we see the world and leave us in a state of perpetual agitation. Our exposure to such stories has unfortunately grown exponentially in recent decades with the creation of 24/7 cable news, the internet and social media. Even in many public places, you now find a TV tuned into the latest crisis. This onslaught of fear stories has become so ubiquitous and so extreme that it's not uncommon for people to say they need to go on a "media diet." In many ways, these modern channels of communication have extended samsara, allowing its tentacles to spread even further into what now feels like a new kind of duality—one offline, the other online.

We're all aware of stories that are capable of disturbing us and taking away our sense of peace. But probably the most negative and harassing of them is the one about our imminent death. This is the one that haunts us on a regular basis; the one that most of us simply choose to ignore because we are unable to process it and come to terms with the fact that someday we will perish and lose everything we've worked so hard to cultivate and obtain. Within our limited perspective, we may try and come to peace with death as just another facet of life, but even that can't hide the threatening suspicion that life is a tall fence with nothing on the other side.

Ultimately, all stories—including the stories surrounding object-oriented happiness and our imminent death—are just more samsara. Both life-affirming and life-negating stories have the potential to bind and keep us stuck in the mud. In *Generation Wealth*, Greenfield's hedge-fund manager, porn star and bus driver are all living toxic dreams they believe will deliver them prolonged happiness. Each of them believes they are making an investment in their future well-being and that no matter how bad it gets the end always justifies the means. While such cases might seem extreme, they only seem so because of the particular stories these individuals chose to pursue. Everyone is pursuing one story or another, whether it be a story about romance, wealth, recognition, or even enlightenment.

Of course, not everyone has ambitions to become a billionaire or be a pop star. For most people the dream is simply keeping up with the Joneses—that is, maintaining an appearance on par with their neighbors, family and friends. Parents will work themselves sick just so they can afford their children the latest iPhone and be able to take them to Disneyland once a year. The result, often, is absent parents and empty marriages as both parents work overtime to try to create another Instagram moment for the family. And if keeping up with Joneses isn't your story, how about keeping up with the Kardashians?

Another example is the gaining popularity of conspiracy theories (more beliefs) and the story that inspires people to spend every last dollar prepping for what is sure to be a civil war or some kind of world-ending event in which survival depends on having lots of firepower and at least five-years'worth of canned goods. And let's not forget the classics like, climbing the corporate ladder, finding one's soul mate, having the perfect marriage, retiring into the sunset, having a special enlightenment experience, and on and on...

Each of these stories carries with it the cause of samsara— that which makes us interpret life in ways that don't always match with reality. It's not until we examine these stories that we come to realize samsara is a kind of neurosis. This neurosis is so common among us that we hardly notice it. Stories can have a viral effect with even entire populations unknowingly becoming intoxicated. History is replete with them.

Fortunately, as beings with free will, we do have the power to change or discard our stories. Just because we have a disposition for a certain kind of fantasy doesn't mean we can't modify the story and our role in it. A first step toward getting out of the mud is making the important realization that there is a way out; that we needn't be held hostage to our conditioning.

Each of us lives in a personal story of our own making. Let's call it "The Story of Me." We do our best to shape the me-story to our liking. In our me-story, exists our house, office, city, country, family, friends, partner, pets, hobbies, memories, likes, dislikes, and acquired objects. However, what doesn't live within our personal story is the truth. The truth is that which cannot be encapsulated or molded to our liking. It's what remains, even after the story has ended. In the end, all stories are mithya— thoughts, ideas, and vehicles to help us obtain what we want in samsara, whether that be purpose, hope, courage, recognition, or power. Stories are also used to help us avoid what we *don't* want in samsara, such as uncertainty, anxiety, and fear of the unknown.

All stories are a practical means for navigating samsara but like everything else, are temporary and not always reliable. What we're really looking for is something constant and dependable; something that can withstand the unseen forces of samsara—even death itself. In order to find it, we must begin by leaving our stories behind, including our most beloved story— The Story of Me.

We get stuck in the mud when we confuse empirical reality (the world of objects) and subjective reality (the personal story about me and the world) with the truth. The truth is that which is real—meaning indivisible, unchanging and always present. Entrapment occurs because we confuse what is apparently real with what is real. In other words, we become trapped because we take that which is of the nature to change, to be something substantial, permanent and reliable.

The ultimate story is The Story of Me because it's that which is nearest and dearest to us. However, this story is limiting and is always changing in ways we have little control over. It's changing because our outer experience changes and because the body changes along with our preferences. To put all our faith in The Story of Me is to set yourself up for constant disappointment as impersonal forces decide what does and doesn't happen to "me."

The Story of Me is really a story of imprisonment made by our identification with our desires and fears. The me-story is the one Arjuna is following in the first chapters of the Bhagavad Gita and it's the one Krishna spends the remaining chapters of the Gita trying to get him out of. Lastly, the me-story falls short because it's time-bound. Knowing I will die someday puts a damper on life with the understanding that when the story ends, I will end with it.

The next chapter will explore how to get out of our stories, including The Story of Me and what it means to be free of them.

Chapter 8

Getting Out

So far, explained in these chapters has been the qualities of samsara that keep us bound and blind to suffering. We have put a microscope to the unseen forces that unconsciously shape our thoughts and actions, and we have shown how binding likes and dislikes get us stuck in the mud. Given this knowledge, along with the will and strength to carry through with our highest objective (freedom), we no longer need to be like an autumn leaf being blown here and there by the winds of fortune, we can begin to live our life consciously and deliberately, understanding the limits of what the world has to offer.

But with this special knowledge, we need a practice. Practice and knowledge work hand-in-hand. A man wishing to be a carpenter may have read many books on carpentry and already have the know-how required to build a house, but without training his hands, the knowledge is of little use. There are many spiritual practices (some more worthy than others) but what most of them boil down to is being able to manage the mind so that the knowledge, once learned, can go in.

If you want to see the truth, you have to first prepare the mind, and if the mind is continuously agitated and made impossible to polish by certain habits, no amount of knowledge is going to stick. So as seekers looking for an escape from samsara, we must first polish the mind and establish a practice or discipline to manage it; because one thing is for sure—we can't get out of samsara by doing the same thing we've always done. If that were the case, we would've found a way out by now. In short, getting out of samsara requires that we recondition the mind in order to modify our habits and entrenched thought patterns. In order to do that, we need to live by certain principles that help cultivate a steady mind and develop our ability to be aware of the various forces that influence our actions.

No other place is the process for gaining freedom from samsara so fully outlined than in the Bhagavad Gita. Given the proper guide—one who has internalized the teachings and has scriptural knowledge—this ancient "user manual" shows the way out.

The Gita is often referred to as both *yoga shastra* (instruction on discipline) and *brahma vidya* (the knowledge of what is). While it's outside the scope of this book to provide a comprehensive review of both, it's worth mentioning how an individual goes about getting out of samsara from a traditional point of view.

The yoga shastra of the Gita covers psychology, meditation, values and devotion, while the brahma vidya covers Self-inquiry (Vedanta). Because it's difficult to gain Self-knowledge without first cleaning up and steadying the mind in preparation for it, both may be viewed as a single system for arriving at freedom (moksha).

The yoga shastra emphasizes two disciplines: *karma yoga* (the practice of proper action, plus proper attitude) and *samadhi yoga* (mind management or the "sattvic-ification" of the mind). The various phases of spiritual practice (*sadhanas*) as described by Krishna in the Gita are important for getting out of samsara because each help prepare the mind in ways that make it less prone to suffering and more accessible to seeing the truth.

Each sadhana is considered a yoga, meaning "to connect" or "unite." A yoga connects the seeker with the sought. Because Vedanta is an advanced stage in the spiritual journey, seekers traditionally begin with karma yoga, progress to samadhi yoga and then lastly, study *jnana yoga* (Vedanta). Each phase has its distinct purpose and is viewed as an essential step (there is no picking and choosing). Jnana yoga is the yoga of Self-knowledge and is marked by its own three phases: listening, reflecting and the assimilation of the teachings. This process from experiential practice (karma yoga and samadhi yoga) to knowledge (jnana yoga) is important for fully integrating the teachings (versus simply gaining an intellectual understanding of them). Students who skip karma yoga and samadhi yoga often find it necessary to go back after learning Self-knowledge in order to better assimilate what they have learned.

Karma yoga is summarized as proper action (karma), plus proper attitude (yoga being attitude in this case). Proper action emphasizes worship of family, wisdom, nature and humanity and the practice of giving more than you take; while *proper attitude* emphasizes mental balance, healthy acceptance, humility and gratitude. Karma yoga is done with the perspective that we are not separate from nature, we *are* nature and for that reason, should be appreciative of whatever nature has to offer and teach us. We realize that while we get to choose our actions, we don't get to choose the results. So, along with a feeling of gratitude is a healthy acceptance of the results.

Ultimately, the aim of karma yoga is to make us more connected with our environment and in harmony with what is. It also has the effect of neutralizing the ego, which is often found lurking behind our binding likes and dislikes. The karma yoga phase of the journey is important for refining our outward behavior and minimizing any conflict with the world. For this reason, karma yoga as a sadhana is viewed as the most pragmatic. It's where we begin our journey away from samsara and what we return to the most when feeling confused or frustrated with what is.

The next phase is samadhi yoga, for managing the mind. The point of managing the mind is not only to take back some control from seductive thoughts, but to be able to create space between them and awareness. Initially, this will require a practice such as meditation which helps one form an objective view regarding mental phenomena, including feelings and emotions. Without developing this capacity to see thoughts as objects, it will be difficult to break free of the confinements of samsara and its ability to keep us under its spell.

Samadhi yoga is traditionally taught using Patanjali's *Eight-Limb Path*. The eight limbs include ethical obligations (*yamas*), observations (*niyamas*), posture (*asana*), sensory inhibition (*pratyahara*), concentration (*dharana*), meditation (*dhyana*), and a special division-less meditative state where the seer, seeing and seen become one (samadhi). While Vedanta doesn't endorse the philosophy of Patanjali's Yoga (capitalized),[1] it does support its teaching of values and mind management as a practice for preparing the mind for Self-knowledge.

A similar and related system for managing the mind that falls outside the Hindu schools of thought (but was nevertheless influenced by them) is Vipassana (insight meditation). Vipassana comes from the Theravada Buddhist tradition and similar to Yoga, encourages virtuous living, concentration practice and the investigation of mental phenomena. The West's modern mindfulness movement came out of Theravada which encourages the yogi to take an objective view regarding one's inner experience. While Vedanta doesn't agree with the Buddhist's description of consciousness, Vipassana is valued for its mindfulness technique and its description of the process of dependent origination (the cause and effect of mental phenomena).[2]

Along with the development of mental discipline, samadhi yoga also includes physical discipline, verbal discipline and sensory discipline. Physical discipline is the discipline of maintaining a healthy body through diet, exercise and rest; verbal discipline is mostly about managing the quantity and quality of one's speech; while sensory discipline is about avoiding the ingestion of unhealthy objects that can pollute the mind and body. All three disciplines are suggested for creating a sense of balance and making intelligent choices versus just letting one's likes and dislikes run the show. Nevertheless, the primary goal of samadhi yoga is paying attention to thought patterns.

Vedanta teacher, Swami Paramarthananda reminds us to:

Watch your thoughts, they become your words
Watch your words, they become your actions
Watch your actions, they become your habits
Watch your habits, they become your character
Watch your character, it becomes your destiny.

In other words, as you think, so shall you become!

Meditation is an excellent sadhana for many reasons— physical health, stress release, developing concentration, sense of composure and well-being, etc., but in regard to getting out of samsara, it's most useful for seeing where our blind spots are. Meditation is like a laboratory where we get to see and find thought patterns under a very effective microscope. This microscope only comes into focus when we relax the body and quiet the mind through concentration. This allows the yogi the space and quality of attention to see mental phenomena objectively.

Meditation is not just about having spiritual experiences (which seldom happen, by the way) but for investigating how our mind is conditioned. This permits us to recognize and strengthen positive patterns such as kindness, gratitude and compassion, and gives us a chance to explore and weaken negative ones such as anxiety, binding desire, fear and hate. The first step to weakening negative mental phenomena is always through the recognition they exist. Once known, like a scientist, one can observe thoughts, feelings and emotions for what they are—inert objects. By creating such a subject-object relationship with thought patterns we can begin to manage and reshape the pliable mind.

To summarize the value of samadhi yoga, Vedanta uses

the illustration given in the Katha Upanishad which compares the physical body to a chariot, the sense organs to its horses, the mind to the reins, and the intellect to the driver. In order to arrive at freedom, we must have a chariot that is in good condition (a healthy body), a driver with intelligence (a good intellect), the reins firmly held (a disciplined mind), and the horses under control (the ability to manage the senses).

Skillful disciplining of the mind, body and intellect in preparation for knowledge is, of course, work. This work is, in part, why so many people find it so difficult to get out of samsara—because it's hard to change one's conditioning. It can take years to steady and purify the mind due to all the "stuff" we've accumulated over a lifetime in the way of attachments and identification with certain thoughts, feelings and emotions. Even Self-realized individuals will continue certain sadhanas such as meditation and guna management in order to maintain a sattvic lifestyle and help clean up certain vasanas that are not easily vanquished.

In many ways, doing the work is about re-learning what it means to be a human being—one that is more aligned with nature and predisposed to happiness. This requires ample amounts of patience and persistence but is really the only valid means if getting out of samsara is what you want.

In contrast to our fast-pace culture, there doesn't exist any "moksha-on-demand" or "liberation drive-thru." Doing the work must be done systematically and with patience as the seeker slowly matures and moves beyond that which obscures them from the truth. This is why life can be so frustrating, because there isn't another way to actual freedom—you either do the work and with some luck, gain knowledge and control, or you settle for samsara's temporary happiness and perpetual suffering.

While having a practice of self-discipline and mind management is important for spiritual progress, equally important is Self-knowledge which constitutes the last phase of sadhana, called jnana yoga. We first purify and steady the mind, making it sattvic, so that we might prepare it for the subtle knowledge[3] that has the potential to ultimately set us free.

Some individuals are satisfied with experience-based spiritual practices such as Buddhism or Yoga. Like all spiritual practitioners, they value a peaceful mind but are looking for liberation through experiences *within samsara*. Others may become dispassionate about the world and in addition to wanting a peaceful and controlled mind, have a burning desire to be free *outside of samsara*. These seekers already have the view that the world is a zero-sum and that no real happiness is to be found within it. They no longer believe in object-oriented joy and if they've been in the spiritual world long enough, have dismissed any and all myths regarding enlightenment. For these seekers, learning to manage the mind isn't enough, what they desire is a reliable process for gaining and retaining direct knowledge.

When we begin to seek freedom from our suffering, we may first try to adjust the world. We might look at our personal relationships, opportunities for personal improvement and make an effort to influence change. When this no longer works, next, we might try to adjust the mind. We might practice meditation for years, trying to make the mind perfectly polished. When that doesn't work, we go back to trying to adjust the world again, and so on. This isn't to discredit attempts to adjust the world or polish the mind. Both of these have their value *within samsara*. The problem with these solutions is that when we try to hold onto the world, we find ourselves in a position of insecurity where there's a solution that works some of the time, but not all of the time. Even the best yogis can't experience perfect samadhi (meditative absorption) all the time due to the changing gunas. In order to be free of this insecurity, we might arrive at the eventual decision that we need to find something else to hold onto.

At this point the seeker might become qualified to learn of the truth about who/what they are. They are willing to throw away all their stories and know reality for what it is. Whether or not to pursue knowledge or just a sense of peace is a question of how much freedom you want—some freedom or total freedom? Vedanta, a proven means for understanding our experience, isn't for everyone. Among other qualifications, it takes a certain maturity, intrepidness and relentless curiosity as you challenge every aspect of the me-story.

Vedanta's teachings are fundamentally liberating and beneficial when taught by a proper teacher and with the proper methodology. If the teachings were not beneficial, few would be interested in them and the tradition would have died out centuries ago (after all, Vedanta's objective is to free one from suffering, not create more of it!). Nevertheless, in spite of its benefits, there are those who approach Vedanta with trepidation. The last thing they want to uncover is the truth about what they are. For them, it's just too scary a proposition—one that should be left to cave-dwelling sages in far-off lands.

Another reason why Vedanta deters so many is because it's what is referred to as "the king of secrets." It's a secret because for most, even once the secret is known, they still are unable to grasp it. It's not for lack of intelligence that most don't grasp it, but because of lack of qualifications. Vedanta defines the qualifications as the "Four Ds": discrimination, dispassion, discipline and desire.

- *Discrimination* represents an individual's ability to differentiate between what's true and what's not—that is, the ability to see things as they are, not as they appear to be.
- *Dispassion* is a result of discrimination. *Dispassion* isn't a "giving up" so much as a "growing out of." When due

to discrimination, objects no longer have the power to hypnotize us, dispassion is our natural response.

- *Discipline* is the mastery of the senses and mind. This is cultivated through practices such as meditation, mindfulness and adhering to a set of values.
- Lastly, *desire* refers to a burning desire for liberation. This desire is not just a middling desire, but one that is impossible to ignore and set aside. The individual, frustrated with worldly experience and multiple failed attempts at finding lasting peace and joy, begins to look inwardly. With their new understanding that true happiness can only to be found within, the desire for liberation (moksha) becomes strong and the seeker, dedicated.

Vedanta defines real happiness as *ananda*. Ananda has nothing to do with worldly happiness or sense pleasures. While sense pleasures might make our experience of being a human stimulating (after all, what would life be without pleasure and pain?), they are, as we've already learned, ultimately unsatisfactory due to their fluctuating and unreliable nature. Because of our ignorance, they also have a tendency to bind and control us in ways that result in suffering. As beings, we need and seek out pleasure but find that our likes and dislikes, if left unattended, have a negative effect. Thus, worldly happiness in the long run, is ultimately deemed insufficient, leaving us feeling unfulfilled and wanting.

In contrast to worldly happiness, ananda is the complete satisfaction gained from realizing I am already whole and complete. This is the happiness that eludes most people for the simple reason that most people don't consider that the happiness they seek is what they already are. It's so simple and obvious, and yet, because we identify with this body-mind-sense complex and not with the essence of our being (awareness), we totally miss it. In this aspect, we are like the fish who needed to be shown what water is.

The Self is actually not teachable because the Self isn't an object of knowledge. In fact, the only way to know the Self is through a process of elimination (see Chapter 3). Vedanta doesn't show the truth because the truth can't be shown. Instead, Vedanta removes ignorance. Thus, if you want to gain knowledge of the Self and ultimately be free of samsara, you must remove your ignorance about who/what you are.

The closest comparison we have to the Self is space. In Vedanta, space is categorized along with the other basic elements, including: air, fire, water, and earth. Just as we ignore space due to our fixation of the objects in it, we ignore awareness because of our fixation on experience. Like space, awareness is all-pervading, formless, accommodating and can never be contaminated. And like space, all objects come out of awareness and eventually, resolve back into it. The only thing more subtle than space is awareness, a.k.a. the Self.

Unlike the body-mind system, the Self is not a part, product, or property of samsara. Samsara has no effect on the Self. Because the Self is unaffected by samsara, it isn't limited by the boundaries of the body and continues to exist even after the body dies. Like electricity that makes the images on a computer screen possible, the Self is that which makes the body, the mind and the entire world possible. It is the light, the source of everything.

The Self isn't nothingness, either. Nothingness doesn't exist because in order to know nothingness, you need awareness. How else would "nothingness" be known? The fear of death and the dreaded "void" is just another story. After all, each night when you fall asleep you blissfully enter the void without complaint, waiting for karma to gently awaken you once again.

The Self is that which is always present during states of waking, dreaming or even while in deep sleep. The I-sense disappears while in deep sleep, but the Self is always present. If it weren't, your physiological systems wouldn't continue to keep you alive while asleep. The only reason we don't recognize awareness while in deep sleep is because the subtle body (the mind/intellect/ego) is inactive and temporarily unable to know objects. No objects = no thoughts = no experience. So, in deep sleep all personhood is gone. During deep sleep we have no idea who/what/where we are (and to think, we spend one-third of our life this way!). We sometimes get a glimpse of this ignorance when we are awoken abruptly, and it takes us a few seconds to re-orient ourselves.

Self-inquiry is important because it releases us from having to claim ownership of the imperfect body-mind and its downward trajectory of growing old, getting sick and dying. It takes what would otherwise be a story with a tragic end (The Story of Me) and makes it just that—another story. If I am awareness (the movie screen), then I must not be the body-mind (the movie) because *I can't be that which is known by me.* Not only does this free me from the burden of being a person (the movie), it also frees me from the feeling of being impoverished, incomplete and limited. In short, it takes away a whole slew of negative psychological tendencies that bring about the conditions of samsara.

Thus, we can conclude:

The remedy for samsara is right knowledge, plus right identification.

"Right knowledge" would be the knowledge that the world, including this body, is not real; while "right identification" would be *I am the Self.*

Having Self-knowledge doesn't mean we no longer experience the world, it means that we no longer take it to be real—including our thoughts, feelings and emotions. The Self-actualized still experience objects but they see them as ephemeral (including the body-mind) and separate from who they really are.[4] As a result, thrown out are all the stories about object-oriented happiness, along with The Story of Me and "my past," "my present" and "my future."

Another benefit is that we no longer suffer from existential issues emanating from the I-sense. We still honor the I-sense but know it to be just another object on the screen of awareness. We still look after the person's well-being but don't rely on objects or ego-gratification for happiness. The world begins to take on the appearance of a funny play and in short, we're able to take it easy knowing that no matter what happens, I (the Self) am always okay.

In the end, liberation (a.k.a. enlightenment) isn't a special event or special state available only to a fortunate few. You already are free and can never be less or more free than you already are. It's only by the magic of maya that you perceive yourself to be otherwise. Which brings us back to where we began with Arjuna on the battlefield...

Arjuna's fight is our fight. It's our fight to understand that which diminishes us in ways we can't see and in ways that inhibit our ability to live peacefully and happily. It's our duty to steady the mind in order to gain the knowledge necessary to combat ignorance so that we may wield it knowing that in the end, our suffering was all just a misunderstanding; that we were always fine and just needed someone to show us what should've always been obvious.

Endnotes

1 Yoga is one of six orthodox schools of Indian thought. While Patanjali didn't establish the school of Yoga, he was a major proponent of the system through his writing of the *Yoga Sutras*.

2 In the West, Yoga has grown to focus primarily on asanas (physical postures) rather than meditation and mind management. This is unfortunate because the actual focus of Yoga is mind management, not postures. In contrast,

Buddhist mindfulness meditation practice—through its numerous teachers, retreat centers, publications and general popularity—has made the techniques for managing the mind more accessible to a Western audience. Just know that mindfulness meditation is a means to an end, not the end. In order to get to the end, you still need Self-knowledge. Nevertheless, Vipassana is a worthwhile practice in preparation for gaining knowledge and maintaining a sattvic mind.

3   The subtle knowledge that comes from jnana yoga (Vedanta) is different than object knowledge because it's the one knowledge that doesn't change and stays true regardless of a person's experience, feelings, beliefs or opinions. It's the one knowledge that even after thrown in a furnace with all other knowledge, still remains. It remains because it reflects the nature of the Self, consciousness—that which is always present and unchanging.

4   Non-dual vision is seeing that objects are in me but that I am not the objects. Again, the analogy is the movie screen and the objects projected on it.

Epilogue

If you talk to an astrobiologist, they will tell you the Earth is habitable due to the Goldilocks Principle—that is, the conditions that make it habitable are not "too this" and not "too that" but just right. If our planet were any closer to the sun, conditions on Earth would be too hot for life to exist. If it were too far, it would be too cold. Not to mention the important role our moon plays, putting the planet at just the right axis and providing just the right gravitational pull to give us seasons and a consistently regulated climate. In short, everything about the Earth's spatial location makes our little blue planet almost impossibly perfect for supporting life. Is it any wonder that some say the world is "intelligently designed"?

According to Vedanta, the human experience is also a result that suggests a remarkable balance and set of circumstances. Let's pretend for a moment that God is a person with likes and dislikes who struggles to make perfect His Creation. After eons of brainstorming ways to provide a thriving habitat for living beings—including microbes, plants, insects and animals—the Great Architect in the Sky gets around to working on His most endeared, intelligent and beloved of all beings—the human being.

God stumbles a few times with how to make the world suitable for man and have God be known by man. On His first try (a bit amateur, really), God makes the world too difficult for man. Because life is too difficult for man, man never learns anything about his relationship with God and about the essence of who he is. So, man continues to make the same dreadful mistakes over and over again and forever lives in darkness.

Another draft makes the world too easy for man. Again, man never learns anything because man has grown too comfortable and like a lazy couch potato, man has no inclination to do any work or gain knowledge.

A third attempt ensures that man is always happy. No matter what man does—gamble away all his money, lose his home, or get his foot snagged in a bear trap—man is always happy! But, here, also, man never learns anything because man doesn't have pain and suffering to show him the difference between what is right or wrong. So, God tries creating man to always be sad instead, but man just ends up depressed and suicidal, and long story short, no one wants to procreate, and the species abruptly ends.

God's final version is man's current situation: a brilliant plan that leaves man ignorant but gives him just enough insight to provide a way out of his suffering. The suffering that man experiences can be boiled down to his feeling a sense of limitation. So, God's clever plan ensures that man is always, consciously or unconsciously, looking for freedom from his limitation. Man seeks food to be free from hunger, shelter to be free from the elements, and a partner to be free from loneliness.

To challenge man and make his search for God and his self more meaningful, God thinks of clever ways to perpetuate and maintain man's ignorance. God employs Maya, the great illusionist, to draw up the functional specifications for samsara—a hypnotic, dualistic world where everything is constantly changing and nothing is what it seems! In samsara, everything feels real and stable, but is not. Like sandcastles at the mercy of the wind and crashing waves, samsara eventually sweeps away everything—objects, people, countries and even entire worlds!

Maya's most clever trick is convincing man that his happiness lies in objects. This causes man to wildly pursue objects and experiences in the belief that they will provide him lasting satisfaction. God hesitates to go along with such an outrageous contrivance, but Maya, being as she is, won't give in.

To make things worse, at the last moment God mistakenly

puts man's vision in backwards (oops!) so that man is always looking outward instead of inward. Due to this oversight man always defines reality by what man thinks he sees and not by what is true.

Nonetheless, out of compassion for man's predicament God writes a sort of manual—a "word mirror"—handed down to man to discover his self and understand his relationship with nature. (It's a bit ironic that God is later surprised to learn that so few ever discover His manual. There are even those who use it to perpetuate more ignorance, fear and misery in the world in order to serve their own selfish desires!)

With man now fully operative, God's Creation is complete. To put His Creation in motion, God turns the whole thing like a child's top which, as we know, spins based on the laws of momentum and balance. God doesn't monitor the situation— deciding when to cause a tsunami in Japan or let a reality TV star win a presidential election—God simply puts it all in motion. When the top winds down, God spins it again, over and over, for eternity. Because of this spinning, man is forever committed to movement or action (karma) and to God's laws (dharma). Man is always testing God's laws, but there is no gaming the system. The top spins, and so with it, man.

* * *

If you read between the lines of this little allegory, what it proposes is that (1) life is a setup to make us (2) understand who we are, so that (3) we can enjoy life in harmony with nature. Observable from this perspective, is the fact that life has its own Goldilocks Principal. Much as Earth is seemingly intelligently designed to support living organisms, Vedanta suggests the world is perfectly balanced to frustrate us into pursuing (and finding) the truth that sets us free.

Unlike animals and other beings, man is blessed with an

intellect. Because we have an intellect, we are able to make choices and needn't mindlessly follow our program as animals do. This allows man to manipulate his surroundings in ways both beneficial and detrimental to his or her well-being. The problem with trying to manipulate the world in ways not harmonious to the laws of Creation is that when you rub against God, God rubs back. Each of us is constrained by the amount of physical, psychological and moral pain we can endure—and for good reason! God has put such rules in place to help manage the situation for the Total. Otherwise, there would be no limit to the amount of physical abuse, psychological exploitation and vileness we might inflict on ourselves and others. The fact that we haven't blown up the world several times over attests to the majority of people feeling at least some apprehension about breaking the rules.

But getting in trouble is only one way the world nudges us toward seeking the truth. Life is a zero-sum game. The rich and famous may have their billions, but they suffer just like everyone else—albeit, in more comfort and style. Everything comes with a price, relationships included. This means whatever we acquire and appreciate, we must protect, maintain, and ensure it doesn't get taken away.

Most of us spend our entire lives chasing objects and experiences only to realize later or perhaps never, that it was all just a "chasing the wind." Only a few (exhausted from the relentless disappointment that samsara never fails to provide) will finally make a break and realize that lasting happiness is not to be found anywhere in the world.

And if that doesn't pop your bubble, there's life's biggest and most cruel joke:

You are going to die someday and lose everything.

Contrary to popular opinion and zealous authoritarian types, the goal of life isn't to accrue as much money and world-dominating power as possible. In the end, you're a loser even if you are perceived a winner. So, really, what's the point of the pursuit?

And don't think that you will live on in the form and actions of your offspring. Your children have their own karma and will not be able to sustain the same power, wealth and superior moral stature in the same way you did. After one or two generations, any power, wealth or rectitude endowed to your offspring will have dissipated. Four generations later, even your name will have been forgotten (sooner if your offspring despised you while you were still alive).

Taking this all into account, it's not hard to see that life is a setup to push us toward moksha—a more perfect understanding of our self. The reason the human experience isn't "too this" nor "too that" is to steer us toward the truth about who we are. Because only by pursuing the truth, do we have a fighting chance at resolving our suffering. Anything else is just a temporary respite.

Nevertheless, we don't need a creationist account of the universe to understand our unique situation. That there is freedom once we understand who we are isn't necessarily due to some divine plan, it's just the result of wanting to not suffer from our ignorance anymore. Life just is, any expression we assign to it is solely ours.

In the end, Vedanta is only showing us—using a very logical and methodical way—what is natural to us. And what is natural doesn't include suffering, because if it did, we wouldn't be spending so much time trying to run away from it. There is nothing "spiritual" about life nudging us toward liberation. It's elementary. Once you read the signs, it's the most obvious thing in the world. It's only our denial, and belief it be otherwise, that keeps us in the dark.

Addendum

Outline of Samsara

Primary definition:

A negative psychological condition brought on by the misinterpretation of reality.

Secondary definitions:

A condition of the mind rooted in ignorance.

A fixation on objects and worldly pursuits in order to feel fulfillment.

A hypnotic, dualistic world where everything is constantly changing and nothing is what it seems.

Entrapment by attachment that leads to sorrow and delusion.

Beliefs associated with:

The belief that I am incomplete and that my happiness is dependent on objects.

The belief that objects are substantial, unchanging, independent and always present.

The belief that I am the doer, a separate entity among other entities.

The ten qualities:

It fools everyone

We are all born into this world hypnotized by the many bright and shiny objects. It takes most of a lifetime to come to terms with the world's false promises and realize it's a zero-sum. When we're young, we only see the benefits of objected-oriented happiness, but as we get older, we begin to realize that everything has both an upside and a downside. As a result, we enjoy objects, relationships and experiences but are no longer fooled into believing they are the endless sources of joy we once believed them to be. Samsara fools everyone only because we are ignorant and unaware of maya's tricks. Some of us wish to get closer to the magician in order to better understand her tricks. While others, for reasons not understood, will choose to stay amused and in the dark, never questioning the grand illusion.

Nothing is what it seems
Nothing is what it seems because it is constantly changing and made of parts. If one had the power to speed up time, one would see that all objects are ephemeral—that is, they are constantly modifying and becoming something else. What we find when we closely examine objects is that they have no substance. All objects are made of other objects, they are composites— assembled aggregates that come together for a brief moment in time. Objects become even more ephemeral when we realize through science or Vedanta that the mind is what constructs our universe. Nothing is what it seems because we don't live in a material universe, we live in a thought universe.

Change is its most prominent attribute
Because of samsara's nature to change, there is nothing for us to grasp onto— leaving us anxious and insecure. This is also one of the key tenets of Buddhism that states all objects are ultimately impermanent, unsatisfactory and not-self. By investing in something that is bound to change, we can never be happy.

Duality is its nature
Duality in this case means the world of opposites. In samsara, there is always an up for every down and a left for every right. You can't have one without the other. There is no winning in samsara. There is no actual losing either—it's a zero sum.

Projection and concealment are its powers
Samsara is like a carnival Fun House of Mirrors where we are constantly fooled by maya's twin powers of projection and

concealment. Maya works by first, hiding the truth and next, by creating the illusion of truth. In samsara, what you see isn't what you get! All objects are insubstantial and disintegrate upon close examination. Even atoms are mostly empty space. Furthermore, our minds—influenced by natural phenomena and our own conditioning—create a personal reality that often doesn't match empirical reality.

It is everything

Samsara includes all objects in the world, including these bodies, minds, thoughts and feelings. What samsara doesn't include is awareness. Awareness is that which lies outside of samsara. Unlike all objects, awareness is that which isn't divisible or dependent. Awareness is not a thing because it's the subject. It's that which all objects resolve back into.

It is not real

Vedanta categorizes all objects as mithya or "apparently real" due to their nature to change and dependency on other objects. In contrast, pure awareness is satya; absolute truth—that which is constant, whole and always present. Samsara is only *apparently* real due to objects' temporary and insubstantial qualities. Objects exist because we experience them, but their ephemeral quality makes them as good as not real. All objects are like a mirage.

There is no "why"

Even the sages are perplexed to why samsara exists. "Why?" is a question that leads nowhere. Maya, in particular, is a mysterious force that our limited intelligence is only able to penetrate so far. As soon as one door is opened in samsara, another appears behind it. It's an infinite regression of doors to be opened. At the surface of life, we can find cause and effect, but at the extreme micro and macro levels there is no cause to be found. There are many theories about why creation exists, but none add up. Like a dream, samsara just happens.

There is no end

There is no end to *samsara* for the same reason there is no end to ignorance. All beings are born ignorant. Our minds are hardwired to see reality in a way that conditions us to pursue that which is pleasant and avoid that which is not. Unfortunately, that which we become attached to binds and causes us suffering. It's not until we learn what we are that we can put an end to our suffering.

There is a way out

Only Self-knowledge can take us out of samsara completely because samsara isn't actually a place, but a mental state. Once we have knowledge of our true Self (whole, complete, infinite, non-dual awareness) we can relax because we can begin to see all objects as apart from me, the Self. Life then becomes just a movie playing on the screen of awareness. Whatever happens in the movie doesn't affect me because I am immutable. Once this knowledge is actualized, life is no longer viewed as a burden, but instead, something to be appreciated.

Vedanta Meditation

Vedanta traditionally recommends sitting meditation as a preliminary discipline that helps prepare the seeker for Self-inquiry. However, within the tradition, meditation can also be used as a method for internalizing the knowledge, once learned. Because the following meditation works through the heart, it helps counter the misconception that Vedanta is only an intellectual exercise. In fact, when properly learned, Vedanta can be personality-changing, leading to a whole new outlook on our experience.

Setting Up

- Find a quiet and clean place to sit. The best time to meditate is in the early morning, before the day leaves traces on the mind that we easily become attached to. Think of meditation as a temporary respite from the world. Meditation done properly is like taking a vacation from samsara.

- Sit comfortably on the floor or in a chair, and in a position that allows you to keep still for several minutes. Put your mind and body on notice that it's not nap time by sitting with a straight back, neither too relaxed, nor too rigid. One should be comfortable but attentive.

- Remember to not take any meditation practice too seriously. There should be a certain lightness to your approach, absent of any tension. Meditation is best approached with the karma yoga attitude, which means that whatever happens as a result of meditating (joy, peace, contentment, insight) is not up to you. Meditation isn't about controlling the mind but rather, setting up the right conditions so the mind can become sattvic. Once this is achieved, our job is to get out of the way. Vedanta meditation, in particular, isn't about having any kind of "enlightenment experience." You already are that which you seek, any joyful experience is just a reflection of the Self. Vedanta meditation is about internalizing what is already known—I am the Self. This is why it's important that one gains Self-knowledge before doing this kind of meditation.

- Root yourself in the body by watching the breath and at the same time relaxing the body. It's not particularly important where you watch the breath. It can be at the tip of the nose or simply watching the rise and fall of the abdomen.

- Next, begin to breathe through the heart-center.

- Start with a simple phrase such as, "I am whole," reciting it slowly with sincerity. Repeat it a few times until you are able to feel it in the heart-center. When the feeling evoked by the phrase begins to fade, try another phrase repeating it slowly and deliberately, such as "I am eternal." Likewise, do the same for one or two more phrases using another adjective to describe the Self. Repeat the cycle of phrases if needed, or just stick with one until it resonates. Another way is to start with a few phrases, "I am..." and then afterward, just repeat the word describing the Self, letting it slowly sink in: "whole," "eternal," "limitless," "infinite," etc. Before long, the words will begin to quietly echo within without any effort.

Other phrases you might try:

I am whole... complete... limitless... unchanging
I am the light... formless... eternal... all-pervading
I am that which is beyond pain... old age... sickness... death
I am that which is beyond birth... death... space... time
I am pure... immutable... happiness... peace

- Continue until there's a feeling of resonance in the mind and throughout the body. This feeling of wholeness and limitlessness is your object of meditation.

Distractions

- Observe any tension created by thoughts that appear and distract you from the object of meditation. Don't push thoughts away. Instead, welcome them, but don't feed them. Take a subjective view knowing they aren't you. Give them space and observe how they rise and fall of their own accord.
- Distractions are simply a reflection of likes and dislikes. Don't pay attention to their content or try to analyze them. Only observe the mechanics of their coming and going. By not paying attention to a thought, it will lose energy and pass away of its own accord.

Hindrances

Pay attention to any hindrances to your meditation in the form of the gunas.

- Rajas = desire, anger, or restlessness
- Tamas = dullness, sleepiness, or doubt

Refrain from liking or disliking any hindrance. With nonjudgmental observation, view the hindrance as just another object on the screen of awareness. Notice how it shows up through any one of the sense doors (sight, sound, odor, taste, touch or thought). Observe how likes and dislikes create a subtle tightening sensation in the mind and in the physical brain.

Manage any hindrances using "the 4 Rs":

Recognize that the mind has drifted away from the object of meditation

Release attachment to the sensation or thought
Relax into any tension felt in the body or mind
Return gently to the object of meditation (e.g., "I am whole, complete, non-dual awareness")

Feel the rise and fall of any tension as likes and dislikes, memories, emotions and thoughts come and go. Make sure to relax again after the release of a thought so that the mind begins to regularly settle back into a state of calm and ease.

As in ordinary life, sometimes unwholesome states arise during meditation.

1. Observe that an unwholesome state has risen
2. Refrain from giving the unwholesome feeling more attention, but don't try to push it away or control it either
3. Replace it with a wholesome feeling (e.g. "I am whole, complete, limitless non-dual awareness.")
4. Stay with that wholesome feeling

Once the mind has become quiet and joy has arisen from the object of meditation, stop any verbalization and just appreciate the peace and sense of contentment (sattva) that was cultivated. Thoughts may still arise, but their staying power is now weak and only appear as blips on the screen of awareness. However, if thoughts do become a problem again, re-establish yourself in the physical body and apply the 4 Rs (recognize, release, relax, return).

Resting in Awareness as Awareness

In the final phase of meditation, all internal verbalization has stopped along with any tension generated from likes and dislikes. This phase is marked by peace and collectedness. Expand your attention to include the space, silence and stillness that surrounds you. Let your body fall away and merge with a feeling of expansion.

Lastly, turn your attention away from the object of meditation and turn it toward that which is aware of the object. Rest in this bare awareness which is a reflection of the Self.

Recommended Resources

The following recommendations are for the benefit of those who would like more structured instruction for getting out of samsara. Vedanta offers an ancient and proven method that includes the steps of listening, reflecting and assimilation of the teachings. All of the resources below are a good starting point for putting you on "the pathless path."

Introduction to Vedanta by Swami Dayananda
Swami Dayananda was an important teacher of the Vedanta tradition who had the ability to communicate the teachings clearly to both Indian and western audiences. In this book, Dayananda introduces man's fundamental problem and the importance of Self-knowledge.

Vedanta: The Big Picture by Swami Paramarthananda
A short and straightforward overview of Vedanta based on talks given by a well-known disciple of Swami Dayananda. Suitable for beginners.

The Essence of Enlightenment by James Swartz
James Swartz is best known for making the Vedanta teachings accessible to a western audience using a no-nonsense style free of spiritual trappings. In this book, he thoughtfully and methodically explains the process for gaining liberation while limiting the use of Sanskrit and esoteric phrases. Suitable for all students of Vedanta.

Vedanta: The Solution to Our Fundamental Problem by D. Venugopal
Venugopal's book is more challenging than the books already mentioned due to its use of Sanskrit and references to the

Upanishads. Nevertheless, it provides a comprehensive review of Advaita Vedanta true to the tradition, along with footnotes for those interested in looking up the source of the teachings. Suitable for intermediate students of Vedanta.

http://ExploreVedanta.com
An excellent online reference for understanding the basic structure of Vedanta's teachings. Recommended for all students of Vedanta.

MANTRA
BOOKS

EASTERN RELIGION & PHILOSOPHY

We publish books on Eastern religions and philosophies. Books
that aim to inform and explore the various traditions that began in
the East and have migrated West.
If you have enjoyed this book, why not tell other readers by
posting a review on your preferred book site.

The Less Dust the More Trust
Participating in The Shamatha Project, Meditation and Science
Adeline van Waning, MD PhD
The inside-story of a woman participating in frontline meditation
research, exploring the interfaces of mind-practice, science and
psychology.
Paperback: 978-1-78099-948-7 ebook: 978-1-78279-657-2

I Know How To Live, I Know How To Die
The Teachings of Dadi Janki: A warm, radical, and life-affirming
view of who we are, where we come from, and what time is calling
us to do
Neville Hodgkinson
Life and death are explored in the context of frontier science and
deep soul awareness.
Paperback: 978-1-78535-013-9 ebook: 978-1-78535-014-6

Living Jainism
An Ethical Science
Aidan Rankin, Kanti V. Mardia
A radical new perspective on science rooted in intuitive awareness
and deductive reasoning.
Paperback: 978-1-78099-912-8 ebook: 978-1-78099-911-1

Ordinary Women, Extraordinary Wisdom
The Feminine Face of Awakening
Rita Marie Robinson
A collection of intimate conversations with female spiritual
teachers who live like ordinary women, but are engaged with their
true natures.
Paperback: 978-1-84694-068-2 ebook: 978-1-78099-908-1

The Way of Nothing
Nothing in the Way
Paramananda Ishaya
A fresh and light-hearted exploration of the amazing reality of
nothingness.
Paperback: 978-1-78279-307-6 ebook: 978-1-78099-840-4

Readers of ebooks can buy or view any of these bestsellers by
clicking on the live link in the title. Most titles are published in
paperback and as an ebook. Paperbacks are available in traditional
bookshops. Both print and ebook formats are available online.

Find more titles and sign up to our readers' newsletter at
http://www.johnhuntpublishing.com/mind-body-spirit.
Follow us on Facebook at https://www.facebook.com/OBooks
and Twitter at https://twitter.com/obooks.